Showing what government actually does is a great offense.

WE Republicans

COPYRIGHT

First published in Texas, 2015

Copyright © Mark D. Robbins

SenatorMark4

ACKNOWLEDGMENTS

A LARGE PART of this book was inspired by the words in the platform of the Republican Party of Texas, a document which represents the collected thoughts of thousands of active Republican partisans in the Greatest Country in the World. As anyone who has been following recent politics knows, Republican elites have not yet quit walking down the path that the Democrats have worn so smooth, even though they crushed the Democrats in the 2014 midterm elections. There are times when our elected conservatives appear to be nothing more than working mules: they follow the row.

As is always true, our families should get most of the credit for anything we do. From feeding and clothing us to being the wall against which we bounce ideas, they are the bedrock of everything we are.

Friends are important for the same reason, but because this is a political wheeze, I'm not going to name those who have taken part; I want to protect them from an internet conflagration if one comes.

I want to mention a few people who I really

admire, who have inspired me at a great distance, and are already out there fighting for us every day: <u>Professor Glenn H. Reynolds</u>, the <u>Ace of Spades blog</u>, <u>Dana Loesch</u>, and, of course, <u>Mark Steyn</u>. They all write so well that I am embarrassed even to have tried. Y'all are wonderful. Churchills.

The strength and power and all
praise belong to Jesus.

Trust in the Lord with all your heart, And lean not on your own understanding;"

Proverbs 3:5

Contents

Preface..7

Introduction..14

Simmering Fear..17

Diss-satisfied..43

Preamble...62

First, the First..67

No More Sausage!...92

Public Broadcasting..103

Vote Your Mind..114

Votes Change Leaders...119

IRS form 1099-GOV...142

YOUR Government Accounts.................................177

United States Sovereignty....................................196

Perfected Preamble...222

Where to now?...225

Appendix 1: Data Library.....................................243

Appendix 2: Story Listing.....................................257

Author Pleadings..259

PREFACE

My daughter has been after me for some time to write down my stories. Age and health have led me to include them here while I am still able. In this text, I have integrated some of my life stories with the changing politics that afflict us all. If you're concerned about wasting your time here, you can limit it to ten minutes by looking at "Coral Snake 1989" on YouTube to see what is coming your way. If that makes you smile even once, then you're likely to enjoy the remainder of this Kindle adventure.

The fuse for this project was ignited because I don't believe in spending money or time on a political party that doesn't represent my beliefs. I know that members of the same party can never agree with each other 100%, but surely we could put a line in the sand that's north of 50%. If we can't agree on 50%, then I'm pretty sure I'm in the wrong crowd. Personally, I am past tired of seeing Republicans talk about how proud they are to be working in a bipartisan fashion, even though we worked to give them the kind of electoral victories and majority that they haven't seen in generations. I am really beginning to believe that the Republican leadership consists of nothing more than Democrats who drink bourbon instead of scotch. The volume imbibed is a mirror image. The burning frustration that they ignored in conservative voters led to this rant. Thanks for nothing!

A growing outcry laments the current crop of politicians as so corrupt that the reins of government could never be returned to the people. Many believe that the only way to correct such a thoroughly sullied system is to start fresh with a third party and fight on the field under a different flag, but when outside candidates start working to install a base sufficient for a third party to appear on every ballot, they learn quickly that every election rule, especially in Texas, is designed to make success highly unlikely. I learned this personally when I ran as a write-in against Senator Cornyn in 2008. He had $10 million dollars to fend me off, and all I had was a new hat and some shiny spurs. I talked to thousands of people in Austin bars, and I met most of George W. Bush's Secret Service detail in Crawford, Texas, but nobody understood the point, which is that EVERY vote should count.

As things stand now, third-party candidates must play by big party rules, pay their fees, and engage in battle on their turf, or they cannot succeed. Engaging in a battle of ideas is worthless when the press has already picked sides, and when the press has picked sides, it pays no attention to lesser known candidates. Ignoring opponents is a proven strategy used by all candidates and the press that supports them, because it means no battle and no risk of defeat. In no other field of life can ignoring one's enemy pay such large dividends.

Even if I had garnered a majority of the votes in the Republican primary, I could not have won. In Texas, write-in votes only count if the candidates in question have registered and paid fees (legally registered, officially sanctioned). If they haven't, the county election judge is *authorized* to throw votes in their favor to the ground. In my opinion, this is worse than a poll tax or a literacy test, because it not only prevents a vote but actually allows a legally cast vote to be discarded. This is surprising only to those who haven't attempted to scale the high walls erected by existing parties to prevent actual citizen participation. Candidates are forced to build huge, costly organizations, pay the fees and guide the petitions in order to scale the walls. If they do not, their voices are discounted...utterly. I personally can't deny that a cowboy candidate with a Star Trek-looking badge was out of the mainstream, but if an idea is good, it is good whether or not it is officially registered.

My wife and I were delegates in Fort Worth's 2014 Republican Party of Texas state convention, and we watched a highly organized group jam their plans onto the proceedings. Again. I was presenting a resolution which I called "Resolution for Public Bank Statements," which would have ensured that the bank statements of government were always visible to the public. I still love the idea, because after working in Texas state government for a few years, I can assure you that *if* officials want to issue something as visible as a $150,000 hush payment, they can hide it in any state agency's books. Maybe I shouldn't say that since I only have experience with four state agencies.

Speaking hypothetically, no number of Freedom of Information Act requests would ever be sufficient to uncover a payoff from financial statements delivered by a state agency. Government has had decades to practice the art of subterfuge with our tax dollars. However the monies are tagged by the agency comptrollers, the bank statements themselves record the actual checks which would give a serious investigator access to something factual immediately. Of course, I am assuming that even a government agency would feel uncomfortable issuing large checks as "Miscellaneous." But now, after watching just about everybody accept what might have been classified email from the Secretary of State's private account, I'm not comfortable assuming much of anything. I am pretty sure that agency officials wouldn't pass cash around from government accounts, but even that belief is no longer 100% firm.

Needless to say, my time and money invested in the 2014 Republican scrum fertilized no fruit dear to my heart. Hotel bills, gasoline, and time spent can now be checked off to lessons learned. Again. Insanity it surely is.

I now believe that the easiest, quickest, and most effective way to take our country back is to take over the party that most closely matches our political desires. This chop of the Republican Party platform is my attempt to redirect the road signs toward the free country to which we have always been entitled.

It is obvious that the far left has taken over the Democrat Party. And the media. And higher education. The liberals' vanguard of professional revolutionaries is in place. They have succeeded in marching through all our institutions, and it is time for those of us who are concerned about it to form a conservative, Constitutionally-inspired vanguard and start our own march through the institutions in a broad counter attack. The major difference between them and us is that they know their people must be led because they don't think for themselves, while we believe that freedom speaks to all people.

I think it's time that we all wear white hats.

I found it disturbing that the Republican Party leadership worked with the lame-duck Democrat leadership to write bills in the closing hours of 2014 before the Democrats were replaced by Republicans. Such complicity is unexpected but never surprising, and it is certainly not a welcome omen of their future actions. Politicians will be politicians, but for elected Republicans to hide bills from the people who gave them the greatest majority they've had in generations is just plain wrong. It is unworthy of anyone who claims to be a public servant.

Unwilling to shrug my shoulders at this, I felt it was time to build a party platform that could generate fundamental change—the kind of transformation that would give America back to its citizens before it is too late. More than a country, a set of borders or a people,

America and its founding documents are an embodiment of the *idea* that all men deserve to be free. Such a notion is rare in human history, and it must often be preserved at some great cost.

"Teaching them to observe all things whatsoever I have commanded you: and, lo, I am with you always, even unto the end of the world. Amen."

Matthew 28:20

INTRODUCTION

The platform of a political party is supposed to allow a reader to discern what that party believes. Of course, no party is monolithic. There is variation within the organization for every single platform plank, and there are always, in any group, those who strongly oppose certain positions which are called into question.

My attempt here is not to deceive anyone into believing that this platform will be without controversy, but rather to offer some out-of-the-box ideas which address issues in a different manner. I'd like to see the Republican Party assume positions that can lead voters to put their faith in another way of doing business. Of course, the inclusion of any of these ideas in a platform approved by thousands has its own strength.

For the longest time I've known that the problem with our political system is not a failure of principle. We have a country founded on the greatest collection of documents ever written about man's striving for freedom. By the grace of God, the greatest group of geniuses in history was in the place where they could write the United States Declaration of Independence and the United States Constitution. The ideals in those documents have inspired Americans—and foreigners—for generations.

Some will say that the failures we are experiencing in our national life are caused by a failure of policy. Various groups and their well-paid consultants are working hard to define policies which divide the spoils of government and distribute them among Congress and the parties. People like Saul Alinsky have become idols for entire generations of students and leaders who believe that everyone has the right to steal rather than accept charity or work below their self-determined worth.

On the other end of the spectrum, groups which couldn't be more opposed to Alinsky's ideology understand its destructive effect on society. That said, I might be out on a limb to think that stealing would be discouraged by our Ivy League elites if they were presented with a well-enunciated, critical justification of Marxist theory. The problem really isn't that conservatives don't believe in charity, but that government's perspective on charity depends on who's holding the gun. When armed men of government have the legal right to take money from its citizens, it is not unreasonable for a free people to demand that the money be spent efficiently, fairly, and transparently. Opposing theft is truly *the issue* of process when "redistributing" any asset.

    The focus of my proposed platform
    will be an attempt to define
    individual **processes** which enhance
    freedom, provide for transparent
    government, and, hopefully, serve
    mankind's insatiable desire for
    individual liberty.

I know with absolute certainty that all of you sense a change in our environment. You don't have to listen to the President of the United States bay on about fundamental transformation to have experiential evidence that things are different. It is said that change brings fear, and President Obama says that fear leads people in the American heartland to cling to God and guns. Personally, I think fear is the only valid emotion when people see a world they have known their entire lives start to crumble in front of their eyes.

Therefore we will not fear, Even though the earth be removed, And though the mountains be carried into the midst of the sea..."

Psalms 46:2

Trusting in God as mountains slide into the sea often seems like the only valid option for me today, but I remember when fear was not part of my life. In the late 1950's before and during first grade at Northwood Elementary in San Antonio, Texas, I detected no fear in parents or friends. Now that I'm Papo to my grandkids, I know that today's parents live in an almost constant state of fear.

My family lived on Chevy Chase Drive in Alamo Heights, right across the street from Northridge Park. If I had been able to capture the thousands of hours of kinetic energy generated on the swings and seesaws, I would be living a different life now. From my earliest memory, after being trained to look both ways before crossing a street, I could go to the park at will and spend the entire day there as long as I answered quickly when I heard Mom shout. It was pure joy. Sure, accidents on playground equipment caused tears, but children were allowed to believe that they could swing completely around the crossbar in those days. There was no division, at least in my community, between places we could and could not go. Playing in the alley all day and having dirt clod fights with all the neighborhood kids was part of growing up.

The interesting thing about those back-alley clod fights is that, when we picked teams, all we cared about was how accurately we could aim. The fact that we could all play for hours in the alley—violent, smiling, and without adult supervision—is something, I believe, that no young adult can fathom today. I think it is beyond their imagination, and I have no trouble believing that they would call the Department of Child Protective Services to effect a rescue if they could witness my childhood. They are the same generation that sees trigger warnings for free speech in college as reasonable, which makes me wonder if free time could easily appear oppressive to them, too. What? Alone with nothing but imagination? Heaven forbid!

My earliest clear memory is of the very first

time I can remember reality shattering fear. I'm not sure how old I was, but I know that my sister, who is just over a year younger than I, was nowhere close by. I can remember the scene like it was yesterday. I was playing at my mother's feet as she sewed on her Singer machine. She leaned over and asked me to please put on some underwear. I argued with her because I was wearing a little American Indian loin cloth, vest, and headdress with a feather, and even at this young age, I knew that Indians did not wear Fruit of the Loom.

The exchange between us continued every few minutes. She would pause her sewing and patiently try to make me put on some underwear, but I had already decided that I wouldn't. At some point I was distracted by a party favor, one of the timeless blowout noisemakers which look exactly the same now as they did then. If you blow on one, a coiled sleeve at the end unrolls while it emits a sound —a great thing to entertain a young child. If you blow hard, it toots and unrolls. Blow a little, and it unrolls a little with a less shrill toot. The best part, if you really get into it, is that after blowing it out and unfurling it, you can suck it back in and it will roll back in with a snap while it sings a half note shriller. I'm sure I'm not the only child who tested this particular feature.

Those who aren't parents might think that I had already driven my Mom around the corner by blowing the noisemaker and refusing to don undies. Trying to focus on labor while a minion at your feet is toot tooting away might seem annoying to the verge of dark confusion, but those of us who have children know that hearing *any* sound from a child, whatever it is, is comforting in a special way...until you hear *that* scream. Or complete silence. Nothing gets a parent running through the house faster than complete silence.

So there was my mother, working steadily at her sewing while I sat at her feet, cross-legged, keeping her company, going toot-tweet-toot-tweet. Then, **THREWffffGEEEEEEEEEET!** Ok, I'm not really sure how the whistle sounded when it flew down my throat, but the instant it lodged in my wind pipe, I knew I had a problem. Every time I inhaled it registered one tone, and every exhaling cry sounded another. I was sobbing in G sharp!

You don't have to be a Ph.D. in sports medicine to know that something is wrong when your every breath is metered by a tweeter. My mother knew instantly, too. She spun on her sewing stool, grabbed me by the shoulders and shook me. (If I had died from tweeter asphyxiation **and** shaken baby...would she ever have gotten out?) The vision of her face and hand coming toward me as I tweeted out my anxiety is burned onto my brain. After shaking me didn't work, she just dug into my mouth and pulled it loose. I've had a pretty big mouth ever since.

Most of us experienced at least one existential fright during childhood, but the point is that we did not live in fear from that day forward. Today, police lecture parents about letting their children walk to the park without adult supervision.

The freedom of my youth in a previous America stands in stark contrast to today's first day at school. At the beginning of first grade, parents take their children to school, and the parents meet the teacher. My mother was no different: she packed my lunch box, made sure I was properly dressed, and walked me to school on my first day. I don't remember her crying, but then she was probably more relieved to have me under a professional's care than the crying moms were.

The contrast lies in the fact that she walked me a long way to school. MapQuest measures the distance about 1.3 miles from my house, but I know it was much closer than that because she took me down our alley and then another alley, so I'm guessing that after those shortcuts, we walked about a mile.

An unusual incident painted that day on my memory. On our trip to school, we walked behind a city grader that was scraping our alley back to level. Later that day on our way home, I was paying attention to everything but the route we were taking, so I happened to notice a small hole in the ground that the grader had uncovered. Like any little boy, I stepped toward the hole to see what it was, but a sudden swarm of bumblebees coming out of it sent my mother and me rapidly toward the house. Like screaming lightening, actually!

I don't know how many times I was stung, but the bee that looked at me eye-to-eye from my nose is the unforgettable one. Unlike honey bees, bumblebees don't lose their stingers when they sting. This one was happy to sink its barb into me as many times as I'd let it, so the memory I have of that incident is a screaming retreat and the sight of that bumble bee loving my nose like Miley Cyrus. I can't explain why I didn't knock it off. Perhaps my mind was flooded with bee venom, but once again I have a distinct memory of my mother's hand coming to my rescue as she plucked the bee from my nose.

I spent the rest of the day lying on the den

floor with a fan blowing over me. I think I cried for a long time. I remember my parents sitting above me at the table, rubbing my injuries with ....butter?... and putting ice on my stings.

I have no memory of my mother ever walking me to school again. For the rest of the year, I walked to school down a couple of alleys for a mile or so, all by myself, and it seemed completely natural. The only lasting effect of the bee bombing is that, to this day, I never approach any hole in the ground boldly.

At my age now, I need to hear directions a couple of times before I swear I know the way, so I have a hard time believing that I was sharp enough at age seven to remember the route to school after being shown only once. On the other hand, it was the younger me. Even more astounding is that I was a first grader, *walking to school alone.* Drive by any elementary school today, and you will see hundreds of cars, traffic monitors and frazzled parents, numbed by the slow-moving traffic in the drive-through lane where they pick up their children. The scenario represents a huge investment in vehicles, gasoline, labor, and **fear**. Why are we living this way?

Of course, fear can be a good teacher. As a teenager, I wanted to prove my bold, brave self to women, so I decided to test and advertise myself by riding bulls. I started with dreams of some young cowgirl wearing sprayed-on jeans and swooning over my championship belt buckle. Heck, I even bought my belt too long so I could fold it down behind the buckle; it was the cowboy code signal that meant I rode bulls. Really! I look back at those things—the decision and the dream—and I wonder how I ever got so sideways with reality.

A friend and I practiced on 55-gallon drums which were hung from trees with a bull rope around them so the other guy could shake it for bucking action. Both of us were from ranch families, so we'd spent most of our young lives around cattle and thought we understood bulls. In short, we respected their power but had no fear of them. We rode at jackpot rodeos and county fairs around the San Antonio area—in Hondo, Victoria, Sabinal, Hearne, Uvalde, Criders, Utopia, Del Rio, and Oak Hill west of Austin. (It's hard to believe that a jackpot rodeo arena stood in what is now the daily traffic jam at the "Y" in Austin.) Anyway, I never won any money. I never won a pie plate belt buckle. I never had a girl moan about my cowboy image, either.

I've noticed that many bull riders are smaller men, and because I'm 6'1", I think I know why. The distance from my crotch (where the bull rope is grasped) to the top of my head is identical to the distance from the bull rope to the spot between the bull's horns. Needless to say, I'm smart enough (now) to realize that bouncing my head off a bull's skull is not the way to win at life, but I did conquer my fear because I was allowed to test myself. When I grew wiser, I left that fear in the past. I wish we could all learn from serious experiences with as few injuries as I suffered. (This would be the perfect place to say I got some sense knocked into my head, wouldn't it?)

"The nations raged, the kingdoms were moved, He uttered his voice, the earth melted."

Psalms 46:6

Today, the fear that I feel when I wake up in the morning is of the sort expressed in this scripture—the kind which suggests to me that everything I know and trust may be melting away. I figure that if the Lord utters a word and melts the earth, we'd better have enough faith to control our fear, because there is absolutely nothing we can do about it.

With the Obama administration running things, checking the daily news has become an adventure in anxiety. Oh, they decided to trade guns to Mexican drug dealers in Operation Fast and Furious? Interesting. The Department of Justice has decided to make excuses for people that assault the police in Ferguson, Missouri? Expected. The purported leaders of the country don't think that American citizens deserve to know what is happening in the halls of power, so a dedicated wannabe President justifies hiding emails on her private server accounts at ClintonEmail.com? What, exactly, is surprising?

It seems that some new insult is perpetrated on the body politic every single day. The next couple of years promise more of the same. The Constitution and the Bill of Rights are under attack daily. The Declaration of Independence seems like an ancient idea now that global, 1984-style monitoring of our every action is not only possible, but actually being installed. President Obama appears to be ruling by decree, and the politicians we have elected seem to agree with his actions once they're sworn into office.

This sort of news has created a new sense of fear among Americans. Many sense that the country is in the proverbial hand basket on its way to hell, and they are afraid that nothing can change its course. Survivalists used to be viewed with the stink-eye, but no longer. Elected delegates to political conventions try to choose the best for this last great Republic, but their efforts are totally wasted. Nothing they say or do has an effect on the political machine. The person who spends time and money in an effort to become a voice for the people is considered by the bureaucracy to be less important than the next wave of Syrian refugees. (I'm sure that no Islamic State agents arrived in the latest bunch, because Obama can be totally trusted to prevent terrorists from infiltrating our homeland. Not!)

I don't think anyone in the United States can convince themselves that we *won't* have domestic battles with Islamic State radicals in the near future. Most of us—Democrats and Republicans alike—pretty much agree that Obama will try to disarm the American citizenry when the battles begin. This has always been the ultimate goal of the Democrat left and the ultimate fear of the Republican right. With one swarm of refugees from an Islamic skirmish, enough middle-eastern men to form a military division of Islamic State fighters will be living among us, and we will certainly be giving them welfare until they are strong and frustrated enough to assault the Great Satan. After all, how many Arabic speakers do you need on your job site? Obama will finally have the crisis he can use to disarm Americans, and the gun-clingers will have been proven right.

We all need to ask, "What should I be doing now?" The example of what happens when a society is forced to rely on guns for legitimacy lives in the homelands of the new Syrian refugees. The example of what to expect is the Mumbai, India massacre in 2008 that lasted four days and left 126 people dead.

Being attacked as a political, religious, or racial target breeds a different kind of fear, and if we let it, the resulting distrust can last a lifetime. I learned this in the sixth grade.

A friend and I had been granted permission to watch a movie at the Aztec Theater in downtown San Antonio. Built in 1926, it was a perfect example of the exotically themed theaters built during the roaring twenties. My mother drove us to the bus stop at North Star Mall, she watched us until we caught the first bus, and then we were on our own. We had some money for tickets and snacks in our pockets, and we were feeling really grown up.

I wish I could remember what the movie was, but the day's events wiped my memory clean of pleasant images. After the movie, my friend and I sat on a curb and discussed the movie as we waited for the bus. It was broad daylight on a Saturday afternoon, with people on the street. We knew that things were taking a turn for the worse when we heard the hocking sound behind us, and we turned in time to catch giant phlegm grenades in our faces. A group of Latino boys, a few years older than we were, encircled us, kicked us in our backs and sides, and covered us with the product of their mucous membranes. I have no idea how many boys assaulted us, but they surrounded us so I'd bet there were at least six.

My friend and I looked at each other from under our arms, both of us almost in tears, whispering whether or not we should make a break for it. At that moment, we heard some yelling back and forth from the alley. All of the boys' attention—and ours—turned to the alley, where a small group of black youngsters were shouting, and in a moment our new friends were off after them. We shot off in the direction of another bus stop where there were more people.

You can guess how many times I've ridden the bus to downtown San Antonio since then, and most of you will be correct. As I read at Small Dead Animals blog; "riding mass transit is like inviting twenty random hitchhikers into your car". Despite changing my willingness to use public transportation, I didn't allow the incident to foster too much fear or hatred. I still ride mass transit like the Washington Metro or the San Francisco BART because the stops are monitored, have controlled access, and not just any random hitchhiker can enter your car there without hundreds being witnesses.

What I'm trying to impart, here, is the sad truth that there are people in the world who will pick you at random from a crowd and attack you simply because you're alive. It happened in 9-11, it happened in Mumbai, and it is happening more and more all over the world. All those parents waiting for their children outside the schools feel it. You feel it, too. We're wise to be wary, but to fear what we can't control is a waste of energy.

A more sinister kind of fear is sheer terror. I've experienced that, too, and it recalibrated my anxiety meter. I'm not talking about the terror in books or television documentaries, but the kind that shakes your body with sobs of dread—a kind of terror that you understand only by living through it.

In November of 1973, when I was in College Station, Texas, attending Texas A&M University—letting that greatness soak in a little—a friend and I decided to do some hunting in west Texas. We left mid-afternoon on a Friday and arrived after midnight at my a ranch south of Fort Stockton. We were met by other hunters who had arrived earlier and had a Wild Turkey start on us, so we caught up as best we could. After a couple of hours' sleep, we were off for a full day of chasing the big mulie.

We saw lots of deer but nothing we needed and after an unsuccessful day of hunting we spent another long night killing Wild Turkeys, sharing our best hunting stories, and racking up a few hours of sleep before doing it all again. My friend and I knew that we had to head back for exams (or something) at school on Monday, but we were delayed: every time we turned a bend and saw a deer, it just wasn't the right one. We eventually headed back, but we didn't hit the road until 7 PM.

After our unsuccessful hunt, we decided to drive back via Del Rio, because we wanted to hunt the long lease—the yellow stripe ranch—on the way back. Road hunting wasn't legal, but it wasn't the huge felony it is now, and we thought we might see something that needed shooting. We were prepared. In those days, a .30-06 Winchester and a .270 Remington Model 700 BDL in the rear window of a pickup was nothing unusual.

We pulled into a truck stop in Sabinal right around midnight. Nobody was in the restaurant except a couple of old white farmers and a group of about six Hispanics sitting in the corner. My buddy sat with his back toward the front door, and I sat facing the cash register. We ordered our cheeseburgers and sat there, numb and silent. Too little sleep, too many miles, and way too much bourbon had taken a toll. Our food arrived and we started eating, grazing like a couple of tired mules with our heads down. Suddenly I noticed that the restaurant was really quiet.

I looked up toward the cash register, and every one of the Hispanics, all about our age, was flipping us the bird. "Hey, check that," I said to my friend under my breath. He glanced over his shoulder and quickly turned back around and said, "Ignore them, and they'll go away." You can imagine how hard I focused on those fries then.

When I heard the front door swing closed, I decided it was safe to look up again. They were gone. I took another bite of burger. Something flashed in the left side of my peripheral vision, and then I saw the knives. The guys who had been at the register were now directly outside our window, flashing knives and laughing. It seemed like there were more of them, too.

This is the stage of the story where stupid intervened. The only thing I can figure is that we knew we were bigger and had loaded rifles in the truck, but neither of us said that, nor did it even cross my mind. We just believed we could escape clean. Young men didn't want to be snitches back then, and I think it's the same today—snitches get stitches—but never mind. (Whenever I tell this story, I share a little wisdom at this point: if you sense danger, call 911. The worst that could happen is that the dispatchers might think you're paranoid. Now, I guess, you might be considered a racist, too, but you'll have to make the choice between racist or dead. Call 911 if you want to keep breathing.)

My buddy jumped up, threw some money on the table and said, "Let's get out of here." I followed him to the door. In the few seconds it took us to get there, a group of the Latinos had positioned themselves outside the front door, right off the sidewalk. Some had knives, one was swinging a chain, and a couple of them were swinging big-buckled belts. It was quite a sight.

To this day I can't imagine how the words escaped from my mouth. "Hey, you haul ass to the truck and I'll keep them busy for a second!"

I said. Really! My friend did exactly as I suggested and left me there, looking at those guys with their knives, chain, and swinging belts. Every one of them was smiling broadly and laughing, tittering amongst themselves in Spanish.

"Weee gonna cut choor guss out, gringo," the ringleader kept saying, and every time he said it, the others laughed louder. I could tell they weren't native English speakers.

My response was, "We didn't do anything to y'all!" I said it twice, but I knew it was pointless after the first time. The second time sounded a little whiny, even to me. I'm sure that their failure to listen wasn't strictly a language barrier; they just wanted to kick some white-boy ass on a slow Sunday night, and there we were. After the second "y'all" I was standing at the passenger side of the truck, yanking the handle and yelling across the truck, "Just get in and unlock the door, Just get in, Just get in!" But to no avail.

The thugs had stopped my friend at his door, and they wouldn't let him turn around to get in. We were surrounded. Were there really fifteen or twenty of them, or was it my imagination? That old cartoon image of the white explorer in the cooking pot, surrounded by dancing natives...we resembled that! My whole body was shaking so hard that I thought I was going to dump in my jeans. The gang was punching me in the back of my head and body slamming me against the truck, and I knew it wouldn't be long before they started trying to let the air out of me with their knives. ASTERISK 1.

At this point I actually accepted that I was about to die, and the most amazing relaxation washed over me. I decided that I wanted to die standing next to my friend, and I pushed my way around the vehicle to be next to him. They shoved me, pushed me and hit me on the back of my head as I slipped some punches, yet I managed to make my way around the truck and stand next to my friend.

The moment I arrived, the leader once more broke new ground. He jumped between us, took my friend's .30-06 and slammed it to the ground in front of us. My buddy picked up his new gun, glanced over it, and said, "You got no business touching my rifle." ASTERISK 2

ASTERISK 3. I was actually surprised that all hell hadn't broken loose, because my buddy had a hair trigger temper, which, for those who don't speak "y'all," means that he could get really angry in a New York second. In this case, though, he remained calm. He placed the gun gently back in the seat of the truck and turned around. Unbelievable! ASTERISK 3.

Then it happened. The ringleader jumped toward us, and using both hands, slapped us both. Hard. I saw the blood flood my friend's eyes. He was starting his windup, and that's the last I saw of him for a while. I grabbed the nearest guy's throat with a death grip and was about half way to a great overhead right, when they swamped me to the ground.

Deep calls to deep in the roar of your waterfalls; all your waves and breakers have swept over me."

Psalms 42:7

I had control of nothing but the guy whose neck was in my left fist. A few of his friends were beating on my arm trying to free him while others kicked me and struck blows with their knuckles and belts. It was like slow motion. I saw each one coming at me and hitting me with total clarity. Oddly, I wasn't aware of any severe pain. ASTERISK 4.

When they freed my only victim from my

grip, I tried to slither under a car, but they grabbed my legs and pulled me out. There are no words to describe what it feels like to be dragged into a crowd that is shouting for murder. "Stark terror" might come close, but it seems a little light to me. An antelope being eaten alive by lions would understand.

As it happens, the way I was dressed affected the end of the story. It had been very cold,—in the 20's—and lucky for us, my buddy and I were wearing winter clothes. I had on cowboy boots and Wrangler jeans, and I topped them with a thick wool shirt and a heavy lined plaid wool coat. Under it all was long thermal underwear, top and bottom. It's not too far fetched to say that I had two inches of wool and synthetic fibers shielding my back.

When the gang of thugs dragged me out from under the car, I was on my back. After they delivered a couple of kicks to my stomach and the side of my head, I rolled over on my stomach to protect my face. When I rolled over, they began hitting my back, and after a few blows I felt cold air on my skin. The thought that I was being stabbed in the kidneys, along with the realization that I had maybe a minute to live, washed over me. ASTERISK 5.

In an instant, I spun in place and saw a man's crotch right in front of me. I hit a nameless pair of gonads with all the power a dying man could muster, and their owner dropped like a sack of potatoes. The momentary hole in the mob was all I needed to free myself, jump into the truck, and retrieve the .30-06. I don't remember anyone being in the way. I don't even remember my feet touching the ground. I jacked the bolt, a shell flew out of it, and now **I was the one yelling murder**. Apparently the language barrier vanished because everyone was running.

At that instant, the ringleader came running past me, and I figured that no one deserved to die more than he did. If I was going to die, I owed it to the world to take him with me. I was leaning into the recoil of that .30-06 and was microseconds from blowing a hole in his back big enough for an elephant to jump through without touching the sides, but my friend jerked the gun out of my hands. (NO asterisk here!)

Instantly I was transported into a grey, safe, warm world and a clear, all-encompassing sensation was that I was safe, loved and protected. I don't know if my eyes were open or closed. I don't know how long I was in this state—seconds or minutes—but for the moment I was outside time and space. Then BOOM! A rifle blast brought me back to the middle of a screaming melee. My buddy had discharged the rifle into the ground to get the thugs' attention. I was sad to be back, especially to this! ASTERISK 6

After the incident, my whole body surged and shook so hard that it was an hour before I could hold a cup of coffee without spilling it. After hours of taking statements from people and discovering some of the knife weilders were on probation, the Latino highway patrol escorted us down the road past the county line, and I'm glad they did, because several members of the mob had been threatening to "get us," and there was a packed carload of characters past the county line.

My mother took a Polaroid image of my back the next morning. Apparently the gang just whipped me with belts. Who knew? We never heard another word from the police about the attack, but we did hear a rumor that the ringleader of the gang eventually stabbed a guy to death in a pool hall. ASTERISK 8.

I have to note, here, that the reason I've included this story has nothing to do with gun control. I fully support the Second Amendment and feel naked without my piece, but neither stance is based on that terrifying night. My purpose in sharing it is to show how the world has changed. Every person involved in that incident, except one, was opposed to murder. Most of the mob just wanted to beat down some white boys, and the white boys just wanted to get away. How different things are now. Racial and religious conflicts are now led by those that won't be satisfied without major blood and destruction.

Today, it is easy for *anyone* to imagine being a death target. In Iraq, weddings have been attacked with grenades by Islamists because: music. Jews have been murdered en masse in France by Islamists while shopping because: Jews. Middle Eastern terrorists have threatened to target Americans because: American. The fact of the matter is that even CNN (April 2015) has noticed that the Allyhoo Uhgbar crowd would be more than happy to throw us into the sea if we say a prayer under our breath and close it with "in Jesus' name."

Every ASTERISK position in the previous story was a place where a person with real or ideological murder in his heart could have taken advantage, but didn't. We were all different then. All of us, even the bullies. Now, the sight of a decapitated body is as common as it was in the dark ages. Burning people at the stake is back with a vengeance. These are all clues—signs that we're entering an age which requires a new ideological approach to problems.

I want to share a couple of words about how hard it is to confront our enemies. Wendy Sherman, as Clinton's policy coordinator for North Korea, was a big part of the North Korean Nuclear Agreement. My view of the results: we feed them, they make nukes. Sherman was Obama's Under Secretary of State for Political Affairs during the Iran deal. That agreement is more direct: we pay them, they make nukes. Both agreements are tagged on Wendy Sherman's résumé. Wendy Sherman. Wendy Sherman. Wendy Sherman. Scary. When government can't be trusted, the power is reserved to the people. When the rule of law fails, the people cling to guns. We still have that freedom in America and THAT is the reason for the Second Amendment.

The next few years may be the last ones in which our republic can be saved by its citizens. When Iran gets the bomb, we'll see changes. When a Middle Eastern party to a nuclear deal can safely assume that chanting "Death to America" will not change the negotiating position of the "American" administration, we have entered a scary new world. The founders of the United States of America had high hopes. They longed for freedom and **individual**

41

rights, and we must fight hard, using the tools of government that they left us, to prevent our grandchildren from having to fight real, bloody battles to win back what was freely passed to us. If we do anything less, the light of Liberty will be nothing but a spark in a howling gale.

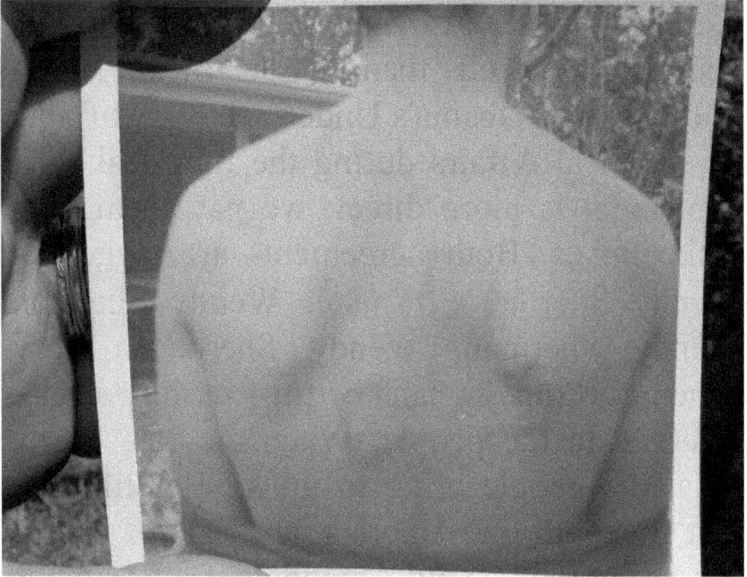

DISS-Satisfied

I've never believed in picking the best electoral candidates from a herd of swine and wasting my vote on them. The party, in my mind, is the machine that carries its principles and processes to the place where they become reality in American life.

I've been personally involved in politics a few times, always as a Republican, and the lessons I learned when I moved among politicians are too numerous and subtle to be listed. I'm beginning to believe that only America's youth can have real enthusiasm for politicians, because principled people old enough to have personal experience with them eventually become disillusioned. In my own experience, getting up close and personal with them made me feel dirty. And certainly stupid.

The place where one encounters government doesn't really matter. The experience is always mind opening, and generally not in a good way. I learned this for myself in 1977.

I had just finished trying for what could have been a B.S. in Microbiology at Texas A&M University. At that time, a Bachelor of

Science required ten hours of a scientific language. French, German, and Russian were my choices, and I selected German. The classes were designed to teach nothing except scientific vocabulary for the purpose of interpreting experimental data in foreign publications. For five days a week, we memorized lists of vocabulary and endured weekly tests that measured our ability to make sense of an experiment.

I just needed those two classes to graduate. Two!

Even now I can't believe I was that stupid. Just as I paid consequences for ignoring the rules for delegate selection, election monitors and signage at the polling place, I didn't pay attention to requirements for graduation and was arrested by reality! The next time A&M offered these scientific language courses was the following fall semester, which meant that I would be another year out for graduation!

Life is full of those places where a comma, a piece of paper, or 75 hours sitting on your butt can make your life ricochet in a different direction. When I learned that I'd have to spend another full year in school, I enlisted in the United States Navy for six years in order to finish my degree, with adventure. I was given the choice to attend any boot camp I wanted, so of course I chose the one that was coed. (Describing how desperate male-female relations can become in stressful situations is simply beyond words, so I'm going to leave that for another time. You wouldn't believe me if I wrote it down here anyway.) So in 1977, there I was at RTC in Orlando, Florida, attending boot camp, and participating in a special program that would allow me to earn a B.S. in Medical Technology from George Washington University in Washington, D.C. It seemed so totally logical and smart. I would finish my degree at a respected university and see the world. Even I could see the advantage of a little discipline in my life. And, remember, the Navy is always by the beach. Pretty smart decision, I told myself.

The first day in boot camp, I awoke at 5 AM to the drill instructor kicking a trashcan down the aisle between the bunks and hollering, "Pop tall!" I can tell you that I felt...disconcerted?...confused? Nope. Stupid, once again. Thinking of adventure and finishing my degree, I had signed up for six years of this. I was starting to believe that the horns of a bull would be bashing me between the eyes for quite a bit more than eight seconds. The thought came to me that adventure, as defined by someone else, just might be different from what I had imagined.

People bring back plenty of stories from boot camp—stories of oppression for oppression's sake and of achievement defying all odds. My own story defines the principles on which military training pivots: follow the rules and obey orders. In minute detail.

"Whoever can be trusted with very little can also be trusted with much, and whoever is dishonest with very little will also be dishonest with much."

Luke 16:10

Having been blessed with lots of college credits, I was sort of special in boot. I was thought so, anyway. I was so special that they conferred a special title on this college boy: they dubbed me Recruit Petty Officer First Class and put me in charge of a company platoon. I don't know if others' experiences mirror my own, but I learned that before I became responsible for others under my authority, I had lived in a very small world. My small world was about to be shaken open.

After our platoon received instructions on the correct performance of our duties—bed-making measurements, shirt folding, and underwear labeling—we quickly had our first inspection. Imagine twenty or thirty bunk beds on each side of a long bay, with a little four-shelf metal rack facing the center of the bay at the end of each bunk. All the newly designated recruit petty officers were assigned bunks at one end of the bay. The rules for the inspection were simple enough: we were to face our lockers, stand at attention, and sound off when approached by the inspecting officer. If we passed, we were allowed to go to parade rest. If we failed, we had to do an about face, turn to the center of the bay and remain at attention until the inspection was over.

Without saying it, we all knew that "fail" probably had some interesting extra-curricular activities associated with it. What we never knew was whether or not group punishments would be dished out. Because those who passed continued to face outward, they didn't know who failed, but they could estimate the number of failures by the volume of shouts. The failures, on the other hand, could look each other in the eyes. Or eye.

I was the first person inspected in my row of bunks. Leadership has its privileges, and I knew that "pass" would mean standing at attention for less time than all the other recruits. Thinking such thoughts about a first inspection is common, but I now know it's also an ignorant pipe dream.

The lieutenant approached my rack. I sounded off, and the inspection began. Each item of clothing was examined for the proper fold. In military boot-campy worlds, everything seems to be measured in 1/8-inch increments, and underwear folding is no exception. I was extremely proud of the quality job I had done folding.

Then it happened. One stupid little instruction, too simple not to remember, too easy to overlook because it went against all previous experience, was going to cost me. Recruits had been told to remove the tags from our underwear and t-shirts. As the inspectors found each of my embarrassing tags, they put on quite a show of name calling, ridiculing the "leader" and the "college boy," and then placing the tags into my mouth so that they hung over my chin. By the time they finished, I had a complete shuffled deck of tags decorating my lower lip. To finish off their performance, they pulled a pair of underwear over my head, making sure that only one eye was exposed and the entire collection of tags remained visible in my mouth.

"Bout face! Ten hut!"

It was as funny looking as you can imagine. I was the first one to fail, and now I would be standing at attention longer than anyone in the company. The amount of pleasure I took from this has rarely been surpassed. You see, when recruits from the other platoon failed and were forced to turn around, there I was in all my tagged glory for them to see. The first guy that failed turned, saw me, and doubled over in laughter. That was worth fifty pushups. Uh huh. "Instructed pushups," if you know what I mean. It was more than a little satisfying to see someone get bounced off the floor for laughing at me.

The best part of this exercise was that nobody could anticipate what they were about to see. Who coulda thunk it? After hearing scores of pushups being ordered, the remaining recruits began preparing themselves to handle the sight of me without degenerating into the laughter that would earn a physical consequence. On the other hand, even those who had done their pushups and had been staring at me for a while could be brought to the floor once again with a wink of my visible eye.

"What are you laughing at, Recruit?"

"He winked at me," they'd say with wasted breath.

"Did you wink at him?" the officer yelled a nanometer from my nose.

Forgive me, Abba... "NO SIR!"

"Are you sure, Recruit?"

"YES SIR! I have to blink now and then!"

"Do NOT let me catch you winking!"

"SIR, NO SIR!"

No duh. Each subsequent set of pushups required a little more bounce-assist from the staff. At the end of it all, I was demoted and "shamed" back to normal recruit status, because it was obvious that college hadn't prepared me—or anyone else—for something like this. And that was the point. Within the boundaries of military discipline, only our leaders and what they demanded of us were important. Status, wealth, race, education...none of it mattered. We were a unit with one job, and in this particular case, the job was de-tagging and folding underwear.

This is, in my mind, is exactly what the founders were trying to emphasize when they wrote the Constitution of the United States. They wanted to derive simple rules of governmental conduct which, if followed assiduously, would apply to all and protect the Americans as a unified nation of free people. A government so constructed would be a government *of the people.* Within the framework of the Constitution, citizens would be able to structure their lives and remain safe in the knowledge that their government was *restrained* from attacking them as individuals or constricting their rights. Each American would be above the government, but no American would be above another under the law.

Reading the previous paragraph makes me shudder. The founding fathers wanted specifically to prevent what has come upon us: this country (and most of humankind) is teeming with people who want to run your life and mine. They want to tell us what to eat, how to spend what we earn, how much of our money they deserve, how to train our children, and what we should or shouldn't read. These people, who believe that the Constitution is a set of suggestions rather than rules of law, are increasingly predominant in the halls of government.

"Woe to you, teachers of the law

and Pharisees, you hypocrites! You clean the outside of the cup and dish, but inside they are full of greed and self-indulgence."

Matthew 23:25

Today, most of us don't relate to the social class that schedules a ritual cleaning of the silverware, but we are directly connected to people who feel that they're above the law. We experience it at every level of every organization, from the boss's son receiving special treatment at the contracting company to high government officials pushing the envelope of privilege. Can you believe that the government allowed Hillary to have a private email system?

In the early 1990's I worked at the Texas Water Development Board, which was my first stint in a state bureaucracy, and the experience awakened me to the reality behind the curtains. While I could talk at length about ghost workers and the misuse of funds, the culture of state government workers alone was beyond shocking.

After years in the system, people become

blind to the familiar, and when they relate their experiences in government, they often sound like fabulists. If I hadn't lived them, I doubt I'd believe them at all.

During the 90's, for example, smoking bans were pretty new, and smokers were loud about it. I think I remember that smoking was banned within twenty feet of an entrance, which was proved by a definite lack of butt cans. I smoked at the time, but I was also trying to be strong, so I always walked the stairs two steps at a time up to the sixth or seventh floor. I'd been doing this since forever.

One day I was returning to work after smoking a cigarette outside. Walking up the stairs as usual, I was met by the Commissioner of the General Land Office (D-TX). He was a man who held a statewide elective office representing the state of Texas, and there he was, coming down the stairs, enriching the atmosphere with a lit cigar in his mouth. I just said hi and continued up the stairs through the cloud, but my face must have given my irritation away because he tucked the stogie behind his hand. The column of cigar smoke floating up the stairwell was more difficult to hide.

It really made me angry. I remembered a janitor who had worked on our floors—a really nice, friendly, mentally-abled guy—who had been fired for smoking a cigarette in a closet, and here was an elected state manager, ignoring the laws completely and arrogantly. I talked to friends about it, but they weren't surprised in the least. I was led to believe that it was no secret, much like Hillary's "secret" email server. Staff apparently knew, too. By this time, I'd worked in state government long enough to know that going through proper channels to complain about something as minor as smoking violations was a waste of time, at least when reporting an elected boss. Still, the thought of him walking around smoking in a state building consumed me. I rolled all kinds of options around in my mind. I decided that I would **not** just walk by him next time, but what was I going to do? Stopping him and verbally confronting him didn't seem like enough. Wasted words. I knew that a written report would be the same as giving someone else a piece of paper to throw away. Direct action was called for.

I eventually decided that if anyone came walking down the stairs with a cigar or cigarette, I would grab it from his hands and put it out on his suit. Sounded just right to me at the time. "Uh, no, officer. I accidentally bumped it into his shoulder." I decided I would not apologize for burning a hole in someone's jacket.

I never had the chance to try it, because soon afterward I took a job in a basement office of the building. (The Lord Jesus was watching over me again.) I still walked up the stairs every once in a while, though, which approached the thrill of hunting!

The cigar smoker is a perfect example of the elected elite, who truly believe that laws are for little people, and when little people like us realize that the normal route for addressing problems means nothing, our personal ideas for solutions float to the top. If laws don't apply to elected officials, why should they apply to us? Authorities' belief that they are above the law is a state of mind which rests on the edge of anarchy. "Occupy _____" is a group that is encouraging us all to stand on that edge. Does anyone really, Really, *REALLY* believe that hundreds of people at the Department of State did **not** know about Hillary's private email server when she was Secretary of State? For years, no one knew? Some of them probably wished they could snuff out a slightly used cigar on her pantsuit.

But I digress. Glenn Harlan Reynolds of Instapundit.com put it best: "But what about the rest of us? If presidents can violate the law, why can't we? It would be a bad thing for the country if Americans started to ask that question" [USA Today, April 19,2015]. .

How much money must a person owe the IRS in order to become an MSNBC commentator? What pay grade must government officials attain before they can rationalize creating their own shadow IT departments? How much justice does a group feel is owed to them before they deem themselves justified in targeting *our* police?

I don't know about anyone else, but I think many Americans are feeling so powerless that they are willing to do nothing until the ship of government somehow rights itself. Hundreds of historical examples show that automatic recovery from chaos is exceedingly rare. Hens' teeth. History also shows that societies in chaos have a common denominator: depravity at the top.

We must strengthen ourselves against the cynicism and passivity which the corruption of our government has engendered, and we must remember that **"Republicans believe that every day is the Fourth of July, but the Democrats believe every day is April 15." (Ronald Reagan)**

When I was young, my family owned a nice little family ranch in Wilson County, Texas. It was a beautiful place: 300 acres in the sandy hills just east of San Antonio, and covered with blackjack, shin, and post oaks. Unlike most of our neighbors, my father had grown up on a real farm, so he plowed most of the trees down in order to plant the land with coastal Bermuda. When the season was good, the ranch was a green postage stamp in the middle of those brown hills, and beautifully decorated with hundreds of Brangus cattle, their black shapes dotting the rolling pastures.

After my freshman year at Texas A&M University, when I was home for the summer, I was put to work helping my father, the ranch manager (I'll call him Curtis) and Curtis's son castrate the herd's young bulls. I think it helps to understand how wonderful I thought I was: I had graduated from high school a year early because I'd gone to summer school for two summers, and when I came home the following summer as an Aggie sophomore at eighteen, I was just too special...my hair, my clothes, everything. (My kids still can't believe that I had hair growing on the top of my head. I can't believe that I thought the center part was stylish.)

Anyhoo, the four of us were working hard. Scores of little black calves, many of them close to 200 pounds, were mingling in the corner of the pen. Curtis and my dad, a veterinarian, manned the table where the business end of Dad's knife made them steers. Curtis's corn-fed high school son (I'll call him Jacob) and I wrangled the unlucky young bulls that starred in the operation. Jacob and I would walk behind the group of calves, and as one of us grabbed and lifted a designee by the hind legs, the other would grab the front legs, and together we swung the calf to the surgical table. He was then cut, doctored, released and herded to another pen. It was a smooth and steady operation.

An hour or so into the process, I had to take a break and have a drink. College boy no have stamina. While I was sitting on the top of the fence with my Dr. Pepper, Jacob kept on working. He wasn't much bigger than I, but he was honed for the job. Grabbing a little bull by the hind legs, he'd swing it onto the table like a sack of potatoes. Dad and Curtis would catch it like pros and continue the job. Their progress seemed faster without my help!

Well, I was so arrogant that I believed watching the three experts was all the training I needed, but I was fixing to get a lesson. When Jacob decided that *he* needed a little break, I told everyone that I'd take over for him now that I knew the "secret." I exercised the kind of wisdom that socialists employ when they analyze a small business: reading about a business is the same as running one. Not.

I wasn't totally unprepared for this exercise. I'd spent thousands of hours tossing hay bales, and quite a distance, too. I'd ridden some rodeo —poorly—but I just knew I had a handle on this calf-throwing thing. I walked up to the first prospect and jerked his feet off the ground. This was the fail, because I didn't know I had to start swinging the calf the moment I lifted his feet. I had the ass of that calf about eyeball level when he started kicking to get free. I learned the hard way that calves can kick really fast when they sense imminent gonadal danger! Before I could even order my hands to LET GO, the young bull drilled a half dozen sharp blows into my chest.

Specialized tasks require experience, and without it, someone usually ends up bruised and hurting. A common problem among the elite left is that most of them have no real world work experience, and their ivory tower regulations governing our lives leave us, the people, bruised and hurting. While we attempt to recover, they write a grant request to finance another study to figure out why their attempt at managing our world failed so horribly. They just know they can get it less wrong the next time.

We already have a proven model for society, structured on the principles of the Constitutional provision giving us freedom to govern our own lives. We don't need to review the failures of other countries to know that America's successes, unparalleled in human history, are proof of the Constitution's soundness. Dealing with politicians and bureaucrats is a mixture of frustration, sharp reversals and occasional sweeping successes; if we are to preserve the best elements of our founding fathers' vision, we have to be organized, determined and prepared for abusive opposition.

"When I was a child, I talked like a child, I thought like a child, I reasoned like a child. When I became a man, I put the ways of childhood behind me."

1 Corinthians 13:11

PREAMBLE

The preamble of a document outlines the document's purpose and underlying philosophy. Judicial courts in the United States have determined, I think I've learned, that actions which violate the structure or ideals of the Constitution's preamble also violate the entire document.

Personally, I'm tired of hearing leftists say that we are obligated to pay welfare because the preamble to the U.S. Constitution cites one of its purposes as promoting the general welfare. As if a farmer in the days of the founders wouldn't look across his musket at someone trying to redistribute the hay he'd gathered with his hand rake.

I think we can disagree about whether or not welfare truly promotes the general welfare, and even about the point at which government's responsibility ends. Most of us Texans, I believe, would agree that government's ultimate responsibility is to create an economic atmosphere that encourages job growth and affords citizens as much freedom as possible. We are tired of government inserting itself into every crevice of our lives in order to determine how much we should contribute to the general welfare. This is the field of electoral politics.

An interesting feature of the preamble to the Republican Party of Texas platform is its effort to give us control of our representatives. **According to Rule #43A**, candidates may have to note their position on no fewer than ten (10) but no more than twenty (20) items in the preamble *before* they can garner any funding from the state party. Plenty of Republican Party of Texas (RPOT) members will read this and say, "How ridiculous! The money flows the other way."

And it's true. Candidates collect most of the money and, if they feel generous, give some back to the party. The cynic in me says this is because buying a vote means buying the voter. If a person running for any political office follows rule #43A but is refused money from his or her party, both the press and the people will look sideways at that candidate. The rule defines the preamble as the place where Texas citizens can apply the most pressure to their representatives. Sure, the candidates can raise money at other barbeques, but if their own party refuses to back them financially, how certain of media support will they be? In Texas, it's called "holding their feet to the fire." Once party funding is denied, we can ask pointed questions. Where is the money coming from, and who is the candidate representing? Whose money is allowing him or her to ignore what the elected representatives and delegates to the convention say is important in an election cycle? I think that holding candidates' feet to the 43A fire would make a difference. No system of checks and balances works perfectly, but we know that corrupt money corrupts absolutely. Do we want George Soros and Middle Eastern money funding our candidates? No. Here's hoping that the people of this country care.

The RPOT platform preamble should be a tool used to chip away at the duplicity of politicians. A spine transplant. A pebble of principle.

The first paragraph of the RPOT preamble reads:

> We STILL hold these truths to be
> self-evident, that all men are created
> equal, that they are endowed by their
> Creator with certain unalienable
> Rights, that among these are Life,
> Liberty and the Pursuit of Happiness.
> The embodiment of the conservative
> dream in America is Texas. Throughout
> the world, people dare to dream of
> freedom and opportunity. The
> Republican Party of Texas
> unequivocally defends that dream. We

strive to preserve the freedom given to us by God, implemented by our Founding Fathers, and embodied in the Constitution. We recognize that the traditional family is the strength of our nation. It is our solemn duty to protect life and develop responsible citizens. We understand that our economic success depends upon free market principles. If we fail to maintain our sovereignty, we risk losing the freedom to live these ideals.

Engaged Republicans generated this paragraph and campaigned to have it accepted at the largest Republican convention in the world. It now lives as a monument, but to what? If the word "Republican" were changed to "Democrat," a huge majority of Texans would support it, so I'd like to believe that the words represent the best of American standards. Unfortunately, party loyalty often comes before an intelligent examination of principle.

"You are the salt of the earth, but if salt has lost its taste, how shall it's saltiness be restored? It is no longer good for anything except to be thrown out and trampled

under people's feet."

Matthew 5:13

In my opinion, we are fast approaching the place where the Republican party—Lincoln's party which freed the slaves, the party which acknowledges a Creator, the party which holds to the idea that government serves the people rather than the other way around—has lost its way and has become little more than an appendage to business and government. Government and business. Indivisible.

I believe we can change things, and I intend to propose changes to the RPOT platform preamble which will fundamentally change the Republican party, and the country, if they are adopted.

First, the First

The First Amendment to the United States Constitution is a dagger to the heart of tyrants, and to prove it we need look no further than the constant efforts of tyrannical governments to restrict it. Clearly, regimes which deny free speech to citizens are not interested in true freedom.

> "Congress shall make no law respecting an establishment of religion, or prohibiting the free exercise thereof; or abridging the freedom of speech, or of the press; or the right of the people peaceably to assemble, and to petition the Government for a redress of grievances."

A country's level of societal success can be predicted accurately simply by checking the degree to which its people support the First Amendment. In 1961 John Kennedy said, "While we shall negotiate freely, we shall not negotiate freedom." He was clearly a different kind of Democrat from the ones in Washington today. Eleanor Roosevelt, another Democrat from the mid twentieth century, is often credited with the successful inclusion of Article 12 of the Universal Declaration of Human

Rights. Adopted by the United Nations in 1948, it states, "No one shall be subjected to arbitrary interference with his privacy, family, home or correspondence, nor to *attacks upon his honor and reputation.* Everyone has the right to the protection of the law against such interference or attacks." (Italics mine.)

These words infuriate me. They sound good on the surface, but they laid the groundwork for encroachment on the First Amendment. As recently as 2011, the United States adopted UN resolution 16/18, which seeks to limit speech viewed as "discriminatory" or which involves "defamation of religion." The resolution was a response to alleged discrimination of Muslims after 9/11 and in the last four years has been revised several times in an effort to make it acceptable to American representatives who are still dedicated to the preservation of the First Amendment. The most recent revision, which *criminalizes* speech that "incites violence against others on the basis of religion, race, or nationality," won US approval even though it indirectly limits speech considered "blasphemous." This is the very type of negotiation that impelled President Kennedy to draw a line in the sand in 1961. The difference between the words of the UN's 1948 Declaration of Human Rights and its most recent resolutions are a glaring example of UN corruption and our government's willingness to subject American freedoms to international diplomacy. Resolution 16/18 codifies support for precepts that are counter to US culture, namely tolerance for **everyone**, even for nations who are intolerant of our Constitutional foundations. It doesn't take much thought to realize that, if enforced, this would be an elephant jumping through the howitzer hole in America's Constitutional rights. Article 12 of the UN's Declaration of Human Rights was probably adopted because, when the world was war-worn after World War II, the US thought that keeping quiet would lead to peace. Surely our acquiescence to this declaration—and the subsequent resolutions to which it led—was just a show vote. We aren't really willing to sacrifice freedom in order to tolerate the intolerant. No way. Right?

Article 12 of the Universal Declaration of Human Rights, alone and without assistance from any other clause in the document, discloses the lie that the United Nation is a bastion of freedom. The ratio of democratic states to monarchies aside, many member nations do not value freedom as America understands it. Worse, the very people who do not believe in freedom and democracy want

Western governments to accept foreign, restrictive laws. Are we really going to allow monarchies and dictatorships to control what we do? Will we really arrive at the place where making fun of the way Bill Clinton used cigars in the White House is legally punishable? Would enforcement of UN declarations and resolutions allow arguments about whether or not President Obama could win the game "Are you Smarter than a Fifth Grader?" after his statement that he had campaigned in all 57 states? (321,000 Google hits, last I checked.) We all know for a fact that Dan Quayle had the same problem spelling "potatoe" that I do, but would betting at bar spelling bees be prevented by a Republican Vice President attempting to implement that particular United Nations article? I'm pretty sure I know how things would turn out if one US court anywhere used that United Nations article to decide a free speech case during the current Obama administration. If the government began protecting politicians' "reputations" (if politicians really have reputations to protect), the level of congestion in the body politic would be mind boggling.

Which reminds me of another Navy story. After completing Navy Hospital Corps School in San Diego, CA, I received orders to the

National Naval Medical Center in Bethesda, MD. After completing in-service training there, I was assigned to the Neurosurgical ICU. Working in any ICU is always stressful, but at that time neurosurgical units in naval hospitals were overflowing with vigorous young men who had lacked wisdom about diving in shallow water or riding motorcycles without a helmet. It was so very sad to see many of these guys literally cut off from the rest of their bodies.

Like all hospital staff before complete P.C.-ization, we had descriptive names for various types of patients. A comatose patient was called a "gork." The system in Bethesda was such that if a gork didn't wake up in six weeks or so, we sent them to the VA if we were short of beds. At times, when it was slow, a patient might stay longer with us, but that was rare if he was not clearly improving.

Mike was a gork. He was one of those patients with whom I spent my extra time because I just sensed that he was "there." I no longer remember why he was in the unit, but I had a sense that he was going to wake up, so I sat with him, talked to him and at times patted him on the cheek. When I could find available time, I spent it talking to Mike during every shift.

One of the preeminent routines in the unit was proper preparation for Grand Rounds every Tuesday morning. The neurosurgeons and all the attached med students and residents would train through like a snake in white: we made sure that all beds had been changed, all wound-care had been freshly completed, and every patient's chart was readily available. The entire medical team would gather around each patient and play twenty questions to ensure that they were thoroughly informed about that case. On good days, I got to tag along.

I remember one particularly interesting day as though it were yesterday. I followed the crowd from bed to bed until my supervisor, HM2 Carla—I forget her last name—called "Petty Officer Robbins!" and motioned for me to come to Mike's bed. It took me only two seconds to realize that we'd be changing Mike's bed and giving him a bed bath, because he was sitting up, writhing like someone who has been immobile for a while, but smiling broadly with the feces on his hands that he was rubbing everywhere—hands, upper body, hair. Everywhere. By the third second I was on my way to fetch everything we'd need to complete the cleanup and bed change. In less than a minute, I was back with my hands full, and Carla was staring at the floor on the far side of the bed.

"Mark," she said in a tone I'd never heard her use before, "you have got to see this."

When I walked around the bed, I was startled, too. Lying there on the floor, looking for all the world like some kind of adobe brick, was Mike's bowel movement. No wonder he was smiling. It is not uncommon for patients on certain drugs to develop fecal impactions, and manually removing them from patients' bowels was a common task for us, but this ... unbelievable! Mike had personally removed something approaching the size of a cinder block from his anus, and I knew that something had been overlooked. The blockage had been fixed, which explained the smile on Mike's face. The cleanup job ordinarily would have taken only one person, but this time it took two because we needed to make ready for rounds quickly.

I view this oddly colored story as a metaphor of the problems we would have if government, with all its guns and agents, was allowed to decide whether or not someone's reputation had been bruised. In Mike's case, someone missed something—either the doctors, nurses, or lowly corpsmen. Three shifts each bore some responsibility, but in the end, the biggest problem was clean up rather than a damaged reputation. If government can criminalize speech that damages a reputation, the victims better have to prove that they had good reputations in the first place, but even then things are not moving in the right direction.

Mark Steyn, a best-selling author and conservative commentator, relates an example of what happens during negotiations between two parties with opposing philosophies. He likens the negotiations to trying to merge a strawberry milkshake with dog poop. If the strawberry milkshake allows even the smallest amount of dog poop in the mix—reaching across the aisle so to speak—the strawberry in the shake loses its essence. It's a great analogy. We are the most free, most successful country in the history of the world, and the United Nations and all the New World Order types are trying to put some foul ingredients into our happy shake. How is it possible to deal with the likes of the Organization of Islamic Cooperation (OIC)? It actually has 57 member states, and their idea of "reputation," which extends to dead people, can ooze redder blood than anything the world has seen since the middle ages. My point is that freedom of speech defines us as a people, and without it we could quickly resemble any number of failed states. A few factions may experience initial joy in trashing a God-given right, but the cleanup is going to be messy. Why do we allow OIC countries to taint our rights? Why aren't we insisting that our proven strawberry shake of free speech be included in the dog poop of OIC countries? Did it work in Germany and Japan after WWII or not?

The story of Mike's gargantuan triumph at the National Naval Medical Center Bethesda didn't end with cleanup. The hospital was an old facility, having been dedicated in 1942 during FDR's administration. It was built with bedpan flushers in the walls, which at the time must have been very modern. Back then, nurses and orderlies took stainless steel bedpans to the wall-mounted flushers, and a bedpan-sized door would open. Once the bedpan was in position, the door was closed, and a rush of steam flushed out the pan when a nurse pulled the handle. Clean and shiny once more. They were no longer used when I worked there, but they provided a perfect place to hide Mike's world-record human...extrusion? Everyone knows that a world record can lead to money, and I was no exception.

As I mentioned earlier, Bethesda was a training hospital. The nurse's station where we all wrote our nursing notes and where the student doctors reviewed the charts just happened to be the same place the…I want to call it something nicer…trophy?…was hiding in the old bedpan flusher in the wall.

I've never been opposed to a bet every now and again, but like most sailors, I knew that suckering an officer into a bet wasn't smart. After all, the people who wear the gold make (and sometimes break) the rules and winning a bet with them often takes second place to their rank. This situation, though, was too rich with possibility to ignore.

Sitting around at the end of a shift, I would casually introduce myself when I heard a group of junior medical officers or medical students talking about various patients, and they'd always have questions about our routine and other minutiae. If I felt that the mark—er, I mean student doctor—seemed reasonable and not too stuck-up, I'd offer the bait. "Oh yeah, Doc," I'd say. "There are real adventures in medicine on a ward like this. We see crazy stuff all the time." That always elicited a response of wanting to know something that might bolster their education in neurosurgery. "Last week—and this is the truth—" I'd say, speaking under my breath, "we had a guy pull a fecal compaction out of his own ass that was this big!"

"I'm sorry, petty officer," was the typical response. "I can't believe that. I'm a 4.0 student at Georgetown medical school, and I assure you that it is humanly impossible to get something that big out of a human rectum without surgery," they would say with complete assurance.

"No kidding, Doc," I'd say, still talking like a golf announcer. "There are witnesses." At this point I always had to make a decision about how far I could go with it.

"I'm sure...confident in fact," they'd say, "and no witness you can provide could convince me."

"Twenty dollars says I can convince you in 30 minutes," I'd reply, raising my voice just enough and leaning in, eye to eye. Part of my technique was to leave them with the idea that I would have to go fetch a witness, perhaps a doctor, which always gave them the chance to say they didn't have time to wait but a few minutes.

I knew that twenty dollars was a small price to pay for education regarding a true miracle of nature. If my plans actually proceeded to this point, the price of learning was never higher than forty dollars, because then it became more than a gag. I kept the numbers low to avoid the possibility of ending up at an Article 15 Captain's mast.

As soon as we shook hands, I'd whip around, open the door of the ancient bedpan flusher and present the clear plastic bag containing the trophy. The typical response went something like "Oh.My.God." times 3, and then "I can't believe you saved it." I don't remember how much I made, but it was less than $200 paid by fewer than ten suckers. Eventually, nervousness about my entrepreneurial endeavor led me to flush it. Literally.

This "poop story" provides an excellent example. Like the student doctors who lost their bets with me, elites never doubt themselves, even when they're dead wrong. Unless someone confronts them, their aggressive, cocksure demeanors will blaze a trail that others might follow. Every Muslim leader in the world agrees that America's freedom of speech needs trimming. Every unelected United Nations bureaucrat thinks things would move more smoothly if United States citizens were just a little bit, teensy bit, hardly-noticeable bit less free to speak their minds. Many of them were trained at elite Ivy League schools alongside our own leaders. While happily taking advantage of America's educational system, which is made possible by two hundred years of free speech, they never doubt their vision or revise their thinking. They simply graduate from Haaavad.

You can have peace. Or you can have freedom. Don't ever count on having both at once.

(Robert A. Heinlein, "Time Enough for Love" 1973)

One of my favorite quotes from a candidate in the 2016 campaign preliminaries comes from Marco Rubio. Speaking in front of the Council on Foreign Relations, he said, **"Mankind remains afflicted, and... its destiny still largely remains in our hands."** I agree. If we want to fulfill our destiny, we have to spread freedom. We have to light fires of freedom that shine out into the darkness of this confused world. The only way we can do that is to insure that tax dollars from free people are not spent on anything that is anathema to freedom, and true freedom **always** depends on free speech.

"For our struggle is not against flesh and blood, but against the rulers, against the authorities, against the powers of this dark world and against the spiritual forces of evil in the heavenly realms."

Ephesians 6:12

How on earth do we do that?

I'm a firm believer in Keep-It-Simple-Stupid. At the end of WWII, the United States imposed and enforced rules that guaranteed freedom to occupied Japan and Germany, both of which are living examples that freedom presages success. Following that example, the Republican Party of Texas should use the golden measure of the First Amendment to beef up the first plank of its platform:

Item 1: "As free Texans, we believe that no United States tax dollars may be invested, granted, loaned, or otherwise expended in any organization or country whose leader(s) do not endorse the First Amendment to the

80

United States Constitution. We believe
that this requirement should be
clearly stated in the Constitution,
and we request that our elected
representative move with utmost
urgency to introduce a Constitutional
Amendment to that effect."

I don't have to preach too much about how Obama and his crew seem to have gone out of their way to arm people who are unfriendly to us and our freedoms. What really strikes home is that his positions are supported by all kinds of elite thinkers. Harvard and Yale-trained attorneys are sitting around tables at this very moment, tossing back fancy whiskey, figuring how to make these deals work, and polishing their arguments for publication in the New York Times. I sometimes worry myself about it late into the night, but then I remember that they were trained up in America and were encouraged to exercise the freedom of speech while being taught the intricate details of our Constitution. This sometimes helps me sleep.

"It is vain for you to rise up early, To sit up late, To eat the bread of sorrows; For so he gives his beloved sleep."

Psalms 127:2

On the other hand, disturbing Yale-trained trophies like Hillary Rodham Clinton actually structure their campaigns around attacking free speech. She recently stated that if she becomes president of the United States, she will impose a litmus test for potential Supreme Court justices by having them promise to reverse Citizens United vs. FEC. This ruling has disturbed the sleep of many leftists, but we never hear them talking about the actual words used in the Citizens United vs. FEC ruling. Here are some of them:

> "If the First Amendment has any force, it prohibits Congress from fining or jailing citizens, or associations of citizens, for *simply engaging in political speech*. When Government seeks to use its full power, including the criminal law, to command where a person may get his or her information or what distrusted source he or she may not hear, it uses censorship to control thought. This is unlawful."

When I read those excerpts, I can believe the words were written by a court that understands America. Hillary, though, wants to reverse the decision containing them. Gathering to petition our government, paint some posters, or heaven forbid, make a political commercial, should not effect the loss of a single Constitutional right, but using that argument on leftists cues them to whine about how corporations are free to destroy our country. (Funny that they never mention unions to me!) The great irony is that leftists are free to say what they like about the system that built this country.

The halls of government are filled with highly trained people who are willing to ignore the brilliance in our founding documents, and they base that willingness on their 'feels.' When the 'feels' of men govern society's issues rather than the rule of law, we will have chaos. The founding fathers of this country worked hard to write legal documents that would guarantee *individual* rights, yet we have entire educational, political, and social organizations dedicating their civic lives to the rights of special interest groups, often at the expense of an individual's freedom to do, think or say what he or she likes.

My point is that we must begin to demand and document people's belief in the First Amendment. It should be the bright line that determines whether or not a person can immigrate, serve in office, or work within the bureaucracy. We have fallen down on this issue so completely that we now have a presidential candidate who openly states that, if elected, she would refuse to appoint a Supreme Court Justice who would not promise to work toward overturning a decision which emphatically endorsed the First Amendment.

Let me share another passage from Citizens United v. FEC:

"The law before us is an outright ban, backed by criminal sanctions. Section 441b makes it a felony for all corporations—including nonprofit advocacy corporations—either to expressly advocate the election or defeat of candidates or to broadcast electioneering communications within 30 days of a primary election and 60 days of a general election. Thus, the following acts would all be felonies under §441b: The Sierra Club runs an ad, within the crucial phase of 60 days before the general election, that exhorts the public to disapprove of a Congressman who favors logging in national forests; the National Rifle Association publishes a book urging the public to vote for the challenger because the incumbent U. S. Senator supports a handgun ban; and the American Civil Liberties Union creates a Web site telling the public to vote for a Presidential candidate in light of that candidate's defense of free speech. *These prohibitions are classic examples of censorship.*" (italics mine)

That seems pretty plain to me. In fact, I have a hard time believing that anyone could be opposed to that reading of the Constitution. But then, I'm me—uneducated, no Yale and probably too much Y'all. The elite seem to believe that they occupy higher ground than the Constitution: when I think of incidents like Harry Reid spouting from the floor of the senate that Romney hadn't paid taxes, then later admitting that it was a political lie, I feel as though governmental integrity is drifting, and not just sideways. When I consider that Hillary wants Citizens United v. FEC reversed, I have no trouble viewing her as a warm, steaming, dense "trophy" like the one Mike presented to us at Bethesda. Flush that!

Enshrining this First Amendment addendum would provide the side benefit of addressing one of the most hypocritical organizations of our day, the United Nations. Wouldn't it be wonderful to see senators demand that the United Nations take a vote on our First Amendment to see where its member nations stand? The theater of such a demand would be worth the price of a year's support. I can envision the Saudi representative shouting down the Iranian delegate, saying that Saudis support the First Amendment much more! The North Koreans would be pounding their desks and demand their right to speak! All would continue to execute their own citizens for speaking against their governments. Theater at its best.

Silly Texans, dreams of spreading freedom are ours alone.

The idea that **no tax dollars** could be spent on any entity that would not stand up for the First Amendment sends chills running up my legs. Having the Egyptian President repeat the First Amendment on video before he gets his billions from the U.S. Treasury would be priceless to me and to his people. The fact of the matter—and every American knows this—is that the United Nations could not exist without the United States and the freedoms enshrined in our Constitution. It is our money and our liberty which allow dictators from all over the world—murderers of gays, persecutors of Christians, mongers of intimidation, eliminators of hope—to convene in a free place and create the illusion that they support democracy and parliamentary procedure. We should make them become YouTube sensations by swearing support for our First Amendment before we share what we've built with patriots' blood.

I know that this is but a dream. If the Republican Party of Texas were to pass such a preamble in their platform and send it out for all Texas Republicans to endorse, what would the Senate say? Would MSNBC have a stroke? Would any politician actually stand up and say that free people should **not** support tyranny in any form? Would our elected representatives ever demand our Secretary of State submit the First Amendment for a vote at the next U.N. General Assembly? It would be a political circus the likes of which we haven't seen in centuries.

Some American businesses throw ethics aside in order to grease the machinery of foreign commerce. They buy politicians who will send money to Egypt, which in turn influences Egypt to buy American goods. They do the same with North Korea, Pakistan, Iran, and any number of nations who are hostile toward the United States.

The only foreign policy that matters to me is the implementation of a rule that tax dollars be sent only to those countries which officially endorse freedom. There is no choosing the correct Muslim insurgency to arm against the

Islamic State. Unlike Obama and his secret Trans-Pacific Partnership, I believe those who receive benefits from free men must announce their concurrence with our First Amendment in public. **For the record.**

YouTube. Popcorn.

Obama's administration announced that they were going to settle 65,000 Syrian refugees in Utah. I'd like to see these "new Americans" on video, supporting our Constitution and having their fingerprints made. My index fingers are on file, as are those of every Texan with a state-issued photo ID card. Shouldn't every alien be subject to the same process? In my simplistic world, a person who lies about supporting freedom in order to enter a free country is deportable. Under Obama's regime, however, it seems that the only people subject to government surveillance are tax-paying citizens.

Obama has been monitoring the Islamic State for some time, perhaps since his times in Indonesia. He's not afraid to say that this conflict we're in with the head-choppers is a generational effort, and I'm sure that everyone agrees. In my opinion, though, we should be fighting the generational war on the basis of freedom, and free men should not support the efforts of those who suppress it. The First Amendment should be the transit for surveying this battlefield.

Anyone who can't agree with this might as well admit that every person—every visitor, every illegal alien, every citizen of the world—should be able to vote in American elections. If our American votes and our culture belong to them, the flip is that every culture, every sharia and every niqab belongs to us. Milton Friedman is famous for saying that we cannot sustain both a welfare state and open borders, but Obama's Democrats are galloping with both, and we seem to be losing democracy in the mix. Ann Coulter described the situation perfectly by titling her book Adios, America.

"The wise inherit honor, but fools get only shame."

Proverbs 3:35

The intrinsic difficulties of open border policies are now obvious, and not just in the United States. In Germany, tens of thousands of Syrian refugees were arriving in Munich, a city of 1.5 million, until the flood of unregistered refugees forced the German government to step back from Europe's Schengen agreement and reintroduce border controls.

Imagine yourself sitting in your apartment while thousands of alien immigrants, in an endless stream, walk down the street chanting. You can guess what they're chanting. If you're unarmed, you cower. I have wondered if that's how the Roman Empire finally fell; only the military and elites were allowed to own weapons, and when the flood of barbarians arrived, the defenders of the realm fled because they were overwhelmed. Why risk their lives trying to beat back the tide? Are we today at the point at which the elites will abandon us to the foreign hordes?

Moving the United Nations headquarters has become one of the most important things we could accomplish. There are plenty of failed nations to choose from, and the Pan African Parliament, from the original home of mankind, could choose one that is centrally located. We would free our country from their sick wits, but also create a safe haven for refugees and migrants. [End sarcasm.] Once relocated, it could be re–forged as a place of freedom (based on the First Amendment) and a land-accessible haven for all migrants. A place of blooming freedom and opportunity could be created out of weeds. If that isn't agreeable to member nations or the Security Council, then the UN can proceed without U.S. dollars. By taking the decision to redirect 10% of our U.N. budget to the Pan African Parliament would generate the kind of interest that might move things forward. And out. Or not. Whatever the outcome, they have to leave.

I don't know about you, but the current assault on the First Amendment makes me believe that the next election is serious. Serious as a heart attack.

No More Sausage!

After several years of slogging through the thousands of pages of Obamacare, we're still trying to figure out what's in it. If the way this complicated, expensive, obtuse legislation was crammed down our throats hasn't taught us a lesson, I'm afraid that our subjection to more like it will continue until individual liberties are snuffed out for good. If corporate/government cronies continue to defy *individual rights* by imposing laws—laws like Obamacare, which the majority of Americans did not want—we will be subjected to much worse consequences than losing our family physicians.

The first time I ever felt like shooting the television was when the McCain-Feingold Campaign Finance Act was announced. There was a bill, Bipartisan Campaign Reform Act of 2002 (BCRA, McCain- Feingold Act, Pub.L. 107-155, 116 Stat. 81, enacted March 27, 2002, H.R. 2356), purportedly written by a 67-year-old Republican and a Harvard-trained Democrat almost twenty years his junior. If anyone in the Republican Party thinks that the interests of the Republicans, or individual rights, were protected by McCain's careful reading of two thousand pages of the junior attorney's work, I have already wasted my time. Those who believe that McCain wrote significant portions of that bill are strangers to reality, but the reason the bill made me want to shoot the television is that McCain, supposedly a Republican, demonstrated a growing tendency of conservative officials to enjoy being led around by the nose, almost like mules tracking behind drivers who control them with a twitch.

The idea of controlling mules with a twitch reminds me of a story that illustrates my point.

When I was stationed at Bethesda Naval Hospital (NNMC) and looking forward to finishing my B.S. in Medical Technology from The George Washington University, I made good friends and enjoyed many adventures all over the East Coast, especially on the eastern shore of Maryland. One weekend the fishermen in our group decided that we should all go camping on the Atlantic shoreline and work at catching some bluefish. My wife (from a previous wife-time) and six or eight others headed out to Assateague Island National Seashore to set up camp. We had a Dodge van set up with shag carpet walls and ceiling, a bed in the back, and a built-in ice box capable of holding a case of beer if packed carefully.

We had packed well and still had beer the first morning. Those of us able to attend to our hooks and lines headed out to the waves at the break of dawn. Running up and down the beach where the seagulls were diving, we cast all the bait we had and caught fish every time. Almost everything we caught was a five-pound bluefish or larger. We'd just toss them on the beach, run to where the seagulls were diving and cast again. If the fishermen's saying "they were biting on bare hooks" ever had any meaning, it was then.

After two hours or so, the fish quit biting. We combed the beach, gathering up our catch, and it was a real chore getting it back to the camp site. The women were already cooking breakfast, and we toasted our success with a few cold ones as we cleaned the fish on one of the picnic tables. Like ghosts materializing, a half dozen of Assateague's wild ponies appeared and gathered at our site. What ensued was one of the strangest things I've ever seen.

As we gutted our fish and threw the offal into the sea grass, the ponies picked it up, carried it off and mouthed it. It appeared that they were eating it. If we walked away from the table, they would rush in and grab bags of chips, then take them a dozen yards away, step on the corner and tear open the bags and devour the contents in a few munches. This was too much! Wild ponies were more adept at opening a bag of chips than the average three year old! In no time, we were putting food on the tables to see exactly how brave they were. As long as we were ten feet away, the ponies were unconcerned by our presence. They loved hot dog buns and chips of all kinds. Pickles not so much.

I decided to see if they would let me feed them. It wasn't twenty minutes before three or four of them were walking right up to me, but they wouldn't quite take food directly out of my hands. I had to drop it in front of them before they would take it. They would put their noses right up to my hand, but any sudden movement would scatter them. In an image you might want to avoid (skip to next paragraph now), there I was in my jean shorts and flip-flops, without a shirt, smoking a cigarette, holding a beer in one hand and feeding wild ponies with the other.

Assateague ponies are small, like Shetland ponies, but without the hay belly that a domesticated pony builds. I was raised around Shetland ponies and learned to ride on one. I knew they were hardheaded, and I had bested their cousins before. In no time at all my mouth was running in overdrive… again.

"Hey you guys", I blurted out. "I bet I can put a twitch on one by hand!"

Those ponies were that unintimidating. I thought I could grab one by a handful of nose and control it. I explained to my friends that you could convince a horse to do almost anything once you applied a twitch. I don't know this for a fact, but I told them you could even castrate a horse while holding him with a twitch. That's how confident I was about my grip. It was beer strength, which we all know is unlimited. As you can guess, it was inevitable that my claims would be tested by reality.

I stayed perfectly still while offering a treat, and one of the braver ponies put himself within reach. In a flash I had a hold on two nostrils and a handful of top lip. In my mind the way this worked was that the pony would buck once as I put the torque to my grip, and then he would stand there stiff-legged while I smiled back at my friends. The reality was different. As soon as I touched him, he grabbed a giant mouthful of my beer gut, and then retreated in four-wheel drive at about 25 mph. I'm just guessing about the speed, because I don't recall that my feet were actually touching the ground enough to gauge the exit speed. I do remember the pony squatting its rear end for traction and kicking up sand. It was a world-record horse/crawdad retreat! After about 20 yards, I was dropped on the beach to the sound of shrill laughter. The ponies stood just a little farther off, and I'm sure they were laughing inside, too.

Of course I deserved it. Huge blood blisters immediately formed on my belly, and I truly believe that a forensic dentist could have identified the pony responsible based on teeth marks alone.

My experience was much like the one that Republicans have been living through for the last few years. We think we have twitched our representatives by showing up to the voting booth in huge majorities and offering financial backing, only to be grabbed in the gut as they sell out to liberal interests again. They ignore the voters' wishes and serve the country up to whomever they desire as long as the citizens continue to pay for it.

"Nevertheless, God was not pleased with most of them; their bodies were scattered in the wilderness."

1 Corinthians 10:5

It used to elevate my blood pressure to see elected Republicans supporting Democrat ideas, but I've grown used to their habit of standing in front of us with attentive looks and serious words and then ignoring us when they get to Washington. I don't want to say that they're Manchurian candidates, but when they believe that the last truly free republic in the world doesn't deserves to see what they're working on before a vote, they are definitely not representing the American people. Otto von Bismarck, the first Chancellor of Germany, is often quoted as saying, "Laws are like sausages,—it is best not to see them being made." Whenever I repeat the quote, I get guffaws and nods of affirmation.

My father-in-law, who celebrated his ninetieth birthday in June of 2015, has been making sausage for most of his life. He's also German, descended from the original German settlers in Fredericksburg, Texas. The story goes that he didn't speak English until he attended first grade, and it still shows when he calls vodka "Wodka." He uses other strange names for things, too. When he's cleaning the meat prior to grinding, he removes all the "leaders" (tendons) and the "silver skin."

I don't help him as much as I should, but I do feel a little better when I do. It's not that I'm scared of being poisoned or something, because he raised nine children to adulthood without any episodes of botulism. It's just that it is interesting—terrifying, really—to watch him make it when I know: 1) his success rate (100%), and 2) how he ignores the FDA rules for meat handling. I know for a fact that he doesn't fetch fresh water for his rag-washing pan when he's cleaning equipment between steps. When I'm present, I always change the rinse water more often, but this difference isn't detectable in the end product at all.

Sausage making and law making are serious business. We might not want to watch either of them happening, but if someone doesn't pay attention to the process, really bad things can happen. Bismarck, who earned the nickname "Iron Chancellor," dismissed the validity of government transparency by likening law making to sausage making. He also installed the first welfare state to earn the support of his working class population. Today, our leaders are installing the first welfare state open to any nationality able to walk across the unprotected southern border. Obama is exercising executive orders as if there were no

Congress of, for and by the people, and the Congress is acting as though they are too busy to care. We *must* start being attentive to the law-making process.

My father-in-law eventually had an accident in which his two middle fingers ended up in the sausage. When he is asked what happened, he says he doesn't know. It became a big joke between us because almost all of my family members are Texas Aggies from Texas A&M University, while all his family are "tea-sips" from Texas University (The University of Texas to some). He had been my serious hunting buddy, and now he sported a body morph that was an exact replica of the tea-sips' hook'em horns hand sign. I often threatened him that if he was to continue hunting with me, we'd just have to take one more finger. His choice, of course.

Oh, and they did throw away that batch. They swear.

Anyway, the point of all my sausage talk is that we need to know what our representatives are doing. Being unaware is not an option, because our freedom is worth much more than a few fingers, a strangely sweet dried sausage, or keeping the family doctor. It is more than just the flavor of society. We cannot trust *any* politician to do the right thing without being constrained by the Constitution, and Obama has proven this over and over again. His replacement, regardless of political party, will be no different if unrestrained. If the First Amendment's right to "petition the Government for a redress of grievances" has any meaning at all, it must allow us to petition our representatives to stop the enactment of laws and treaties that we abhor. If not, then the people effectively assume a defensive crouch that cannot be sustained by any society for long. Our representatives today do "to us" instead of "for us," and then wonder why we don't respect them.

Of the people, by the people, **for the people**.

Item 2: "As informed and literate citizens, we are entitled to have our desires included as input during the construction of laws governing our society. Therefore, candidates for every office will actively represent our demand for a Constitutional amendment stating that no law can be put to a vote or signed by an executive until its complete and unchanged text has been released on the internet for seven (7) business days."

I can hear the whining now. "But no sausage will be made if the people know all the ingredients prior to shipping. People won't buy it if they actually know what's in it." (And the problem is…?) "Sausage is an integral part of our society's customs, and we shouldn't tamper with the process after hundreds of years of success."

Hand them some crackers and cheese to go with their whine. The Trans-Pacific Partnership will ship even more American sausage once we see what's in it. Double pinky swear. Government is out of control, and if we the people don't grab the reins, the acceleration of law being passed without our consent, approval, or even being read by our representatives will continue.

"What sorrow awaits those who try to hide their plans from the LORD, who do their evil deeds in the dark! "The LORD can't see us," they say. "He doesn't know what's going on!"

Isaiah 29:15

Public Broadcasting

It is clear, at least to me, that when the public pays for something, then the pubic owns it or a part of it, which also means that the public should be able to use it. Arguing the validity of the "tragedy of the commons" is less important than thinking about what is shaping the commons of our society today.

Andrew Breitbart famously said that politics is downstream of culture. For instance, when the leftists in Hollywood introduce a story about happy gay parents, then we all supposedly understand why the idea of a gay scoutmaster must then be discussed, because… fairness. Having a gay scoutmaster and a gay scout troop isn't enough: the entire culture of teenaged boys must be exposed to Dan Savage. The <u>University of Oregon pays paid Dan Savage $24,000</u> to talk with students about gay sexual fetishes. I have been waiting with anxious anticipation for a reality show featuring a girl scout troop sponsored by a T.U. fraternity. You know it's just around the corner at this rate of change.

"What sorrow awaits the leaders

of my people--the shepherds of my sheep--for they have destroyed and scattered the very ones they were expected to care for," says the LORD.

Jeremiah 23:1

Discussions about slippery slopes abound, but no one is documenting them to determine how right or wrong they've been. With New York City now underwater and billions of people having died from starvation.... Oh, wait. That was ABC's 2008 prediction of global warming and how the world would appear in 2015. Guess not. (Too bad we don't give grades to experts.) Sarcasm aside, who determines what we invest in our own culture, and why do so many buy into undocumented theories, predictions and philosophies? We may not have experienced any global warming during the lives of current high schoolers, but we live in the middle of a red hot culture which results from upstream dripping that is weaving a very different society from the one that most of us want.

We can pick almost any area of our culture that was once considered relatively sacred, or at least deserving of respect, and find government money undermining it. The National Endowment for the Arts had a budget approaching $150 million in 2015. That's a big pile of money—so big that watching how it is spent is tricky—but a look at NEA's history hints that some pretty serious ideologues and bozos decide where that money ends up.

Back in 1986, the uproar caused by Andres Serrano and his "Immersion (Piss Christ)" agitated some political leaders enough to actually reduce the NEA's budget by tens of thousands of dollars. I have read (and we all know that the internet doesn't lie) that Serrano was paid about $20,000 of public monies for pissing in a cup, plonking a crucifix into it, and then photographing and printing an image of it. (I've wondered if the money was for prostate work to clear his mind.) While I would never think of creating such an image, I find it easy to see the broad acceptance of Dan Savage as a downstream result of it.

It is not unreasonable to insist that the names of people who authorize the expenditure of our money on such things be public knowledge. Images of Serrano's work on the internet makes me wonder exactly what encouraged a bureaucrat to sign off on a proposal to fund this type of "art." Personally, I have a very hard time figuring out what made the artist. "Immersion (Piss Christ)" eventually sold for more than $200,000, and I can't find a single reference to his repayment to citizens of the United States for that stake in his future. Surely this is the sort of problem that **WE can remedy.**

"For our struggle is not against flesh and blood, but against the rulers, against the authorities, against the powers of this dark world and against the spiritual forces of evil in the heavenly realms."

Ephesians 6:12

Yet another cultural pollutant is the Corporation for Public Broadcasting and its $450 million. Politicians knew when they started this empire that there would undoubtedly be partisan attempts at propagandizing the public, and in order to control it, the politicians agreed that the President can never appoint more than five of the nine members from the same party (Public Broadcasting Act of 1967).

I originally planned to write about the clear leftist bias in programs funded by the CPB. I spent some time, looking here and snooping there, in an attempt to figure out how much money Bill Moyers, among others, was worth. This man, who served as Press Secretary and Chief of Staff for President Johnson, has a different recollection from the one of the Church Committee in 1975, which remembers that he ordered FBI investigations of his political opponents. Moyers has also done very well in *publicly funded* broadcasting. The left hasn't changed much in 50 years, has it?

To this day, when something causes me to look around and wonder if it's a bad dream, I can smell the odor of tortured tires. Years ago, I went to a friend's home early on a Saturday to pick up some metal cabinets like the ones found in the storage rooms of every government building. My friend had been to an auction, and I was taking advantage of his good fortune. I picked up three cabinets and put them in the back of my 4-door, 4x4 pickup. Because the cabinets were too long for the truck's bed, I had to leave the tailgate down, but I was sure I had them tied down safely.

As I went south on Interstate 35 through downtown Austin, I kept an eye on the rear view mirror, looking for any movement that meant my knots were failing. I didn't believe they would, because I had learned knots in the good old Boy Scouts of America. Americans in general don't know how to tie stuff up, but I do. Anyway, I watched my load as I traveled on I-35's upper deck. Traffic was moving well—nice spacing but dense—at close to 70 mph. As I was merging again with traffic (2 lanes ground and two lanes elevated) I heard the panic-causing sound of screeching tires behind me. The noise went round and round, ricocheting off the sides of buildings and barriers. I looked in the side mirrors and didn't see anything. I checked the back, and my load was perfect.

"I'm good!" my internal monitor said, and then, right after that thought…*nudge.* Not like a real bump or anything. It was more like hitting an extra large reflector that they use to delineate road lanes.

Instantly, in traffic, my truck and I were perpendicular to the direction I had been traveling. I didn't have time to steer into a skid; I was there. I distinctly remember the sensation of smoke billowing out from both ends of the truck, but I don't think I checked my mirrors. I was frozen, sliding sideways, with a little drift causing the car to head toward the concrete median divider on the left. I don't remember taking my foot off the gas, and I definitely didn't hit the brakes. As I neared the wall, still at speed, I could see the eyes of drivers in oncoming traffic just like a cartoon. Huge eyes.

I know the slide lasted more than a couple of seconds, but it might as well have been an hour, because I slid a long ways. The instant I was relocated in traffic and saw the eyes of the drivers on the other side of the highway, I became conscious of something a friend once told me. "Mark," Mike had said, "when you die, you're going to see it coming."

As I was sliding toward the concrete median, I knew I was going to flip over it into oncoming traffic. It was going to be Marky-parts everywhere, probably mixed with innocent victims.

I hit the concrete hard enough to smash the front of my truck and reposition the engine as though it preferred the front seat. I didn't flip over into oncoming traffic! In the first seconds of breathing life again, my teeth were humming like a dentist's drill. Real fear sings inside you in a way that once you feel it, you know, but you can't describe it adequately to anyone else.

On the second breath, I started taking in my surroundings again. Every car that went by on the other side was gawking at me. My rear end was sticking into traffic, and even though cars behind me were slowing and stacking, they were moving around me. My load was still intact!

"Trust in him at all times, you people; pour out your hearts to him, for God is our refuge."

Psalms 62:8

I looked to my right as I tried to open the door, and there, sticking out of my truck, was a large Mercedes sedan. The driver, a young man with long blond hair, was crawling out of the vehicle. He had t-boned me almost exactly in the middle of my truck, and his Mercedes had been made much shorter. His roof was smashed down the same height the entire length of the vehicle.

"Are you OK?" I yelled at him. He nodded back.

I got out of the truck—I don't remember how—and we talked about what happened.

Short version: semi turned into this lane, Mercedes trapped under trailer spinning around, spun out from under truck, bounced off median divider into me. I do believe in miracles, but to this day it's hard to believe that no one was seriously injured. The highway was thick with cars traveling at speed, yet no one was harmed and only two vehicles were damaged. The driver of the semi had pulled over a mile down the road, and he stopped only because somebody flagged him down. He was as unaware of what had happened as I was.

The story has an important moral: dangerous situations can overtake us in an instant, and surviving one miraculously does not mean that we will survive another one. Disasters create final demarcations between the past and the present and for that reason we must avoid them if we see them coming.

We can't count on miracles. In order to make sure that our society isn't careening toward disaster, we need to set government up to run on automatic pilot with as little bureaucratic interference as possible. Obama and whoever is next will be intent on avoiding Constitutional limitations, so we need to ensure that our nation runs on law, not on the men and women who populate government offices at any given time. Just as the First Amendment has been the gold standard for freedom, transparency should be the gold standard for public funding. Much like sausage making, funding upstream arts and media will put garbage on our plates if we don't involve ourselves in the process.

Item 3: "As intelligent citizens, Texans demand copyright in every cultural venture funded by taxpayers' dollars."

While a demand for copyright is sure to stir up opposition in "educated" circles, it will also instigate a much needed argument about what government's role should be in every part of society. It is settled law that an employer owns the employees' copyrights and patents. Remembering that government is supposed to work for us rather than the other way around, it is time that we demand the same from government. We should absolutely have ownership in the arts and media funded with our tax dollars. When schools justify the use of public funding to familiarize students with ass adventures, we have very much lost our way.

Michelle Obama wants free museums, so I can see her supporting this, too. Symphonies, museums, PBS, and every other **cultural** sources funded by government are owned by The People.

Vote your Mind

Since before the time of the Alamo, we Texans have been "deciders," a people who take pride in our ability to think through problems and assume the heroic stance of a people willing to fight for freedom. We demand no less from our elected representatives. We have no need of elected officials who refuse to take a strong position on any particular piece of legislation.

I am more than tired of President "Present" steering our great country. There seems to be no crisis in the world that he doesn't avoid completely, yet he is still bracketed among leaders by his adoring press. When he occasionally decides to take a stand, it's unclear what in the world he's attempting to change. As a senator, he rarely voted on anything that touched on being controversial, which should make it clear to everyone that he is not now, nor has he ever been a leader. We are now living through the rabble-rousing proof of it.

Playing the safe game, saving a political paycheck with a phony baloney lack of backbone, deciding **not** to decide—all are easy, and all are signs of a complete and total lack of leadership. *Anyone* can fill an empty chair. A vote of "present" requires nothing—not intelligence, not experience, not attention to detail, not even literacy. Babes and fools are capable of it.

"Such a person is double-minded and unstable in all they do."

James 1:8

Claiming that the political process sometimes requires a "Present" vote doesn't work for me anymore, because I'm not interested in saving the skin of some money-grubbing non-decider. Not taking a stand gets us nowhere, at least nowhere I want to go. Those that are voting are the ones that are steering the ship of state, and where they're rowing is not where I think we need to be.

"For you have spent enough time in the past doing what pagans choose to do-living in debauchery, lust, drunkenness, orgies, carousing and detestable idolatry."

1 Peter 3:4

If our elected leaders refuse to tell us whether a piece of legislation either is to their liking or is pretty much crap, then we are poorly represented. WE have to make a stand. It is really that simple.

In 1966 Batman appeared as a television series, and all the kids in my neighborhood met at a particular neighbor's house every week to watch it. It sounds unbelievable now, but this was the only family among us who owned a color television set. My parents were adamant that we didn't need color to watch television; they were raised on radio, so I guess color didn't matter much to them. In any case, every kid in the neighborhood was in front of the color TV 30 minutes early and enjoying their popcorn by the time the first "BIFF" hit the screen in glorious color.

As you know, Batman is not a normal superhero. He is just a regular guy with enough money to experiment with all kinds of cool technology. Because he isn't endowed with super powers, all of us kids sitting in front of that television got the message that, given the right motivation, skills, and investment of time and money, we could become heroes, too.

When you're eleven, that's about all the encouragement you need to start working on hero status.

The simplest tool in Batman's collection was the Bat Rope. He kept it in his utility belt, and it would pop out, bat-shaped grappling head and all, whenever he needed it. When Batman wanted to climb something, he'd just whip out the bat rope and toss it perfectly into some window or tree, and once it was lodged tightly, he'd start climbing. It wasn't hours after seeing him use it that a bunch of us tried out our Bat-line by tying rope to a stick and climbing into the trees in the vacant lot behind the house. To this day I'm thankful that we lived in central Texas when we were Bat Trainees, because none of the scrub oaks were over twenty feet high.

Of course, I was the one hogging our new Bat-line. It seemed reasonable to me because I'd supplied the rope, approved the stick for the grappling end, and was quicker at climbing trees. I was about 15 feet up in the tree when I finally snagged the rope in another tree about fifteen feet away, and immediately afterward I learned that swinging from a rope that would reach the ground was exactly like jumping out of a tree without a parachute. I didn't cry. I was eleven. People were watching. I knew that I'd just missed seriously breaking something, so Mom didn't learn about it until much later. I did learn a little something about geometry, though, and never made that mistake again.

Voting neither for nor against proposed legislation is no safer for our society than jumping from a tall building without a parachute. No one is spared from a hard landing. Representatives who refuse to do it show absolutely no interest in clarifying, improving or re-engineering the system. Politicians who have made a mistake by voting on a piece of legislation with poor public support have effectively generated collateral damage, and they will work much harder on behalf of their constituents to correct the error than politicians who laugh because they stayed completely out of it. We need to pressure politicians to suit up for the game that governs our lives and make it hard for them to stand in the sidelines.

Item 4: No elected Republican will vote "PRESENT" on any legislation and will do everything in their power to implement rules of law to disallow such votes.

Votes Change Leadership

The very idea of a republic gives its citizens both the right and the ability to replace a political team that represents them poorly with one that more closely represents their views. As Republicans, we've learned that this is more easily said than done. Even those who regularly vote Republican are starting to call it the "Stupid Party," which means that the grass roots of conservative support are starting to whither. Too many of us are wondering if our leaders are truly capable of manning our government's steering wheel or are even trustworthy.

When I was in high school and college, my rodeo buddy and I had lots of adventures running into roadside ditches. It always started the same way.

"I'm think I'm going to cut across here", he'd say, "I can make it."

What that announcement really meant was that he was wrangling irritation about having to haul a forty foot gooseneck trailer full of hay or cattle, and he thought he could shave a little time off the chore. My answer was always the same.

"NO! DON'T DO IT!" I'd say at elevated decibels. "If you get us stuck again, I'm not helping. NOT, NOT, NOT HELPNG!"

Then he'd try it. Then I'd help. I had enough self-respect to want to stay out of the mud, so at that point, my turn to drive was guaranteed.

On one occasion, we were working on his uncle's ranch, which covered about two thousand acres north of San Marcos as well as some acreage directly across Interstate 35. The job environment was fascinating and exciting. It was a feeder bull operation which generally kept close to 2,000 bulls on the property in various paddocks. At that time, ranchers could make money just by feeding 400 to 600-pound bulls until they weighed 1,100 to 1,200 pounds, because grain costs were low enough to turn a profit.

Some of you may wonder why anyone would want to feed out bulls. The reason is that bull meat is the leanest beef, and lean beef holds more water when it is processed and canned. When you eat Vienna sausage, you're probably munching bull.

Every afternoon we'd round up about fifty bulls, bring them up to the barn and pen them up so they'd be ready to sell at auction that night. That sounds easy, but bulls don't herd like a bunch of cows. They act more like teenage boys, and there were plenty of disagreements and bullfights among them as we moved them up the fence lines. The ranch was home to a couple of really smart dogs that helped us hold them together as a herd, and they weren't afraid to grab an ear or a nose to make sure an errant bull knew the game. More than once those dogs jerked me off my feet as I walked by a truck they were sleeping under; I don't know whether they'd had a bad dream or just mistook me for a bull. I liked them when I was on horseback, but not so much when I wasn't. Unpredictable, just like the bulls.

"Then I was senseless and ignorant; I was like a beast before You."

Psalms 73:22

The flip side of the operation was that all the bulls sold that night were replaced by younger, lighter bulls which were purchased at the same auction. We'd load the new ones up and bring them back at night, and then put them in pens so they could be worked in the morning. The work made for really good sleep, which was necessary, because mornings started before sunrise.

Any job gets slick when you start a few score times before the sun comes up. In the pre-dawn, we'd load about ten bulls into the narrow chute leading up to the squeeze chute and place posts between the chutes so that they couldn't back out. A bull would be prodded into the chute, and when its head cleared the end of it, we'd clamp its neck down and put the side squeeze on. If the bull had large horns, we tipped them so they wouldn't injure other bulls in fights, but smaller horns were just cut off.

Each cowhand around the chute was assigned specific duties. On my side of the chute, the first task was loading a huge wormer lozenge, nearly the size of a man's fist, into a dispenser which looked like a giant, 18-inch syringe, and delivering the medication deep down the bull's throat. Then I'd move to its hip and vaccinate it with three or four syringes as fast as possible. At this stage, my buddy would hit it with the branding iron, so there was some bawling and bucking going on. We'd give each bull a quick look-over for injuries that might need doctoring, and finally, with our pocket knives, we'd cut a pattern into their ears and tails that marked them as members of this batch. When we were done, we'd set them loose in the pen right behind us, the same pen they shared with our workstation.

I learned something there that has stayed with me to this day. I learned quickly that if I heard a thumping noise coming up from behind, the wise thing was to start climbing whatever was handy. If and when I was safe, only then should I turn around to see what had made me feel threatened. In those days, I usually turned around to see a young bull, wide-eyed and head up, blowing from nostrils as big as my face, and not infrequently, spewing fountains of blood from his used-to-be horns. Working a set of pens designed that way made me quick on my feet.

I tell you this not because I think we're being blind-sided by our Republican leaders (we're not far off, though), but to lay the groundwork for what's next. These were not contented cows waiting to be put out to pasture. They were an agitated set of young bulls, some of which wanted nothing more than to turn something or someone into a wet puddle in the Texas dirt.

I remember one group of bulls that was going to be delivered to the pasture on the other side of IH 35. The gate was directly across the highway, but the route was about a five mile trip. We would have to go north on the IH 35 access road for a few miles, cross over the divided highway, and then come back down the other side of the highway to the gate. My buddy chose this particular day to drive across ditches and medians to—you know—save a little time.

"I SWEAR I'M NOT HELPING YOU IF YOU GET US STUCK!" I said, apparently talking emphatically to myself.

That 1-ton dually 4x4, complete with 40 feet of gooseneck trailer packed with unhappy beeves, met the earth at its axles. And there we were, spanning the median and almost perpendicular to I-35's traffic flow. We might as well have been turtles on a fence post. Fortunately, the traffic back then was not like it is now: there was actual space between the cars.

"Better to meet a bear robbed of her cubs than a fool bent on folly."

Proverbs 17:22

My buddy called for a wrecker, the wrecker

called for the Department of Public Safety. DPS blocked traffic going south, hooked up the wrecker and failed to budge the rig. They called for another wrecker. After about an hour, the second wrecker arrived, and DPS once again blocked the southbound lanes, then hooked both wreckers, and the rig sighed and bawled. I hadn't left my seat in the truck the entire time. I was totally surprised by how well everybody was taking the delay but I have to admit that I was hiding my face in shame, pretending to read. People were getting out of their cars, playing Frisbee, drinking sodas and laughing. Laughing really hard. And pointing. Hooting.

After disconnecting the wreckers for the second time, DPS let traffic go again and made plans with my buddy.

"Run up to the barn with me. We're going to get the horses," he said to me. I just shook my head. No way was I going to face whatever was waiting at ranch headquarters when this story was told. My friend's uncle was not shy about saying that I wouldn't amount to beans because I was associated with so many "incidents." Like this one. Heck, I could imagine him pointing to us and yelling "Sic 'em!" to those dogs, and I knew who would lose that contest. This was one time when I wanted to stay strictly true to my declaration not to help. I waited in the truck.

Within the hour he was back with horses. DPS stopped traffic on IH 35 (Stopped.traffic.on.IH35. Wow!). We unloaded the bulls and chased them into the pasture across the highway. I think they wanted to be back on grass again, because we could usually count on a couple of bulls in any bunch to have their own mind, but this time they all trotted right through the gate with no trouble. A single wrecker pulled the rig free, and in a less than 15 minutes, traffic was flowing again. The DPS didn't even write my buddy a ticket. Of course, back then the traffic didn't back up for miles as it would now. I think everybody was just tickled to see the last great cattle drive on IH 35!

"Professing to be wise, they became fools..."

Romans 1:22

If we Texans allow our elected representatives to take shortcuts which endanger our safety and our future, we have no one to blame but ourselves. Part of their job is to understand that the legal route of sticking to constituents' desires takes longer than caving to political pressure. How many times should we fill Republican quivers with votes, only to watch our elected officials defer to special interest groups? We shouldn't at all, which is exactly why We Republicans of Texas are searching for a way to hold our representatives accountable to us. Accountability is a high expectation during a time when Obama and his entire administration lie ... er ... are not completely forthcoming about their plans for our future. John Gruber, a key architect of Obamacare, said that a "lack of transparency is a huge political advantage. And, you know, call it the stupidity of the American voter or whatever, but that was basically really, really critical to get it to pass. I wish we could make it all transparent, but I'd rather have this law than not. Yeah, I wish there were things we could change, but I'd rather have this law than not." Hearing this blatant admission that politicians did not want Americans to know what was being forced upon them should invigorate our desire to fight their deceptions. We can only fight with troops whom we can trust to make the trip for us without taking shortcuts. The American people deserve better. The greatest country in the history of the world does not deserve to be purposely steered into a ditch.

My grandfather Robbins, the son and grandson of blacksmiths, was also a blacksmith in Coolidge, Texas. Coolidge was (and is) a very small town. When my siblings and I stayed with our grandparents for a couple of weeks each summer, one of the high points of the visit was when our Papo hitched up his buckboard wagon with two mules and drove us around town. The town was so small that we could park the wagon in the middle of the road, go inside and have an ice cream cone, or maybe pigs' feet with saltines, and not worry about our ride being stolen or getting in anyone's way. If the mules lay down and "heyaaaa!" wouldn't get them up, Papo would grab hold of an ear and give it a bite, which made them jump to the ready every time. When I think about that, I wonder if we should notch the Republican leadership's ears or just leave them with teeth marks.

"Men of low degree are only vanity and men of rank are a lie; In the balances they go up; They are together lighter than breath."

Psalms 62:9

Coolidge, Texas was a farming town so small that the graduating high school class was big if it had six kids, but war heroes, flying aces, and solid citizens were spawned there. My father told me stories about harvesting sorghum in his youth—how mules pulled a long, narrow sled between the rows, and how he walked along, cut the heads off the grain with a pocket knife and dropped them into the sled. Two hectares was a good day. He and his brother would compete for the record. That's the kind of work Americans did to build this country, and they were so proud of their freedom that they named their children to honor it. My grandfather was Jacob Isaac Robbins. A great uncle, whom I never met, was General George Washington Robbins. A cousin told me that they called him Uncle General. No flag burners in that crowd.

I remember something else about Coolidge that represents what the ideals of America used to be. In that little town, there were three brick churches. In a town where people struggled to feed their families, they found resources sufficient to erect and dedicate buildings in which they worshiped and gave thanks to God. This being in small town America, the churches were literally across the street from each other, and in a community full of hard men who could pull your arms out of their sockets and make you eat them if they wanted to, all the townspeople worshiped God in their own way, smiling and nodding to each other as they entered or exited the churches.

Contrast that with the Muslim minority that Obama and the elite of both parties are importing to flavor our melting pot. In Islam, the "religion of peace," is there anything like such tolerance? Is there one place in the entire billion-person Muslim world where a Shiite mosque is located across the street from a Sunni mosque? Or better yet, could a Christian church thrive within view of a Sunni mosque? Freedom of religion is one of the most crucially fundamental ideals which make America unique. It is being challenged by Muslim ideology, and the American leadership in both parties seems to want more of it.

Like so much that we like to look back on, that old brick, un-air conditioned church where my grandparents and their sons worshiped is now gone. Whether from diminished population, lack of interest or a caved-in infrastructure, this monument to people who loved God despite barely making ends meet is gone. Hopefully, America will reawaken before our society's foundational precepts disappear, too.

God.Bless.America.

One of my earliest memories of Coolidge included a lesson about work that I didn't grasp until much later. I must have been about four, and I was apparently visiting without my sister. I don't remember much about my stay except one thing: I was standing in the kitchen, and my Mamo was giving my Papo some pretty serious grief. I wasn't frightened, but I knew it was something pretty serious. "You're supposed to be spending time with Mark," she'd say.

"They're paying a dime a pound, Douglas," he responded. (That's right; my middle name is for my grandmother, Douglas.) Even then, I knew what dimes were, and I liked them. After some more of this back and forth, Papo and I headed out together. We arrived at a cotton field near town where we were given a long burlap bag, and Papo went to work picking cotton. He went up and down the rows, bare-handed as I remember, and I was riding on the bag. I remember the crowd as a mix of black and white. I'm thinking it must have been a pretty even mix; otherwise I would have noticed. Or maybe it wasn't noteworthy at my age. My most vivid memory of the day is riding on the bag and watching the tall plants go by. It was tall cotton. Giant, in fact. I remember happy shouts of encouragement and challenges as the bags were dumped and the tally noted. It was a cheerful cotton picking competition. The only thing I can figure is that cash was dear in a small farming town, and everybody appreciated the opportunity to earn a little.

I'm tempted to believe that historians have purposely forgotten about the generations of white Americans who also raised their families by picking cotton. They have openly written that Cortez found Native Americans growing cotton and soon afterward introduced slavery to the continent, but they rarely inform readers that most people in rural areas worked with their hands to make a living. Many, many white families lived as tenant farmers and provided for their children by picking cotton. The Dust Bowl forced tens of thousands of those workers to move to California, where they did the same kind of work on cotton, vegetable, and fruit farms. "Migrant workers" had an entirely different meaning then: most were white American citizens. Despite this historical fact, the only mention of white cotton pickers I've heard in the last thirty years was dialogue in a video history mention of Johnny Cash and his family.

"Laziness casts into a deep sleep,

And an idle man will suffer hunger."

Proverbs 19:15

I am lucky enough to remember my grandfather pulling me on a gunny sack while he picked cotton, and I'm honored to come from such sturdy stock. He and others like him were the backbone of our country, the hardworking citizens who built America, which is the world's last bastion of individual freedom. When we decide that politicians are not protecting that freedom, we have the right to eject them at the next election. Even if we do, we have no guarantee that they will pay any attention to the cotton picking voters. American workers go about their jobs, day in and day out, but even if they turn out at the polls in unprecedented numbers to fire politicians who have failed to represent them, the object of their ire remains on the job for months, collecting a paycheck, and retains the power to pass laws which disregard the people.

Imagine, for a moment, that federal regulations which roughly match a lame duck Congress were imposed on private business.

Workers who were not performing would be retained on the payroll for ninety days. They would continue in their current positions and ranks within the corporation, and though they were officially "Retired on Active Duty," they would have the right to attend every organizational meeting. Does that sound right, or even resemble reality? No? Well, it's exactly what happens in government after an election.

Times have changed since the early days of cotton picking. I used to be amused when Papo drove down the road at 60 mph. He'd mumble under his breath regularly, "A mile a minute." He was astounded by a mile a minute. Now, we fly down highways at much higher speeds, and 60 mph seems slow. Let us never take for granted what astonished the world of our founding fathers: freedom of religion, individual liberty, and the freedom to criticize government without fear of reprisal. We owe it to ourselves and to the country to voice our opinions on the sad state of the Republican Party.

When I was in elementary school, some of the boys started challenging each other, as boys tend to do. As I remember the evolution, we began with the little paper bowls that were used in the cafeteria to serve our food. We took two of them out to the playground, then captured a bee or two between the bowls, and ran around pretending to shave with an electric razor. If the razor went quiet, we'd shake the bee trap to fire it up again. After the bees had too much concussion to be fun anymore, we caught some fresh ones.

Of course, such tame antics didn't keep us entertained for long. We progressed from capturing lone bees to conquering entire hives. When I think about it now, I have to wonder if genetics play a part in the kind of hijinks a person invents. My dad and Uncle Jay apparently used ping pong paddles to stir up bumble bee nests. Dad said it would keep you quick on your feet. The two of them probably put quite the hurt on the town's bumble bee population ... that was before television, and Coolidge has never been known as a happening place. I'm even wondering if my first-grade experience with bumblebees wasn't some sort of karmic payback.

Eventually, my school chums and I graduated to squaring off on Apache red wasps. Apache reds are aggressive. An innocent man can be walking down the sidewalk within ten yards of a nest, and if one of the sisters takes a disliking to his aftershave or shirt color, she'll hit him straight from the nest like a red laser beam. If he responds by flailing and screaming instead of a rapid retreat, airborne support is sure to follow along. (As I mentioned before, it's better to retreat immediately than to analyze the situation!)

In those days, anyone walking through the woods had a good chance of coming across an Apache wasp nest almost as big as a garbage can lid. I haven't seen one like that in decades, but when the boys and I were designing our rites of passage, there was one giant wasp nest that kept our attention for several weeks during school recess. Our elementary school was pretty much out in the sticks, and they were Texas sticks, so no one paid attention to little boys throwing rocks. If we'd been throwing them at each other, someone would have intervened.

The rules of the game went like this: 1) walk up close to a nest; 2) throw rocks at the nest to stir it up; 3) the one who stayed the longest and closest to the swarm of angry red wasps was the winner. As I remember it, two of us were always the winners. Me and my buddy. We knew that the smell of fear would get us carpeted with wasps, so we remained perfectly still, perfectly calm, and perfectly relaxed. The swarm would surround us, but only a few bees would usually land on us. We became expert at letting wasps land on us without reacting,

136

unless something tipped the scales. When an anxious wasp is inspecting a victim, the first sign of its distress is twitching antennae. By the time the twitching gets to its abdomen, a sting is imminent. Most of the time the wasps just wandered around on our shirts or jeans, and then they'd go flying off in search of the enemy. If they got twitchy, we'd move as slow as molasses and set up the thump. We'd wait until they turned directly into position, and a quick thump between their eyes would send them to wasp heaven. Usually. At least often. And as friends, my buddy and I protected each other from the wasps that were out of reach on our heads or the back of our necks.

"When he returned later to take her, he turned aside to look at the carcass of the lion; and behold, a swarm of bees and honey were in the body of the lion."

Judges 14:8

One thing I know from those encounters is that challenging bees without special equipment means learning to deal with stings.

My buddy and I were stung often, but it didn't dampen our enthusiasm for the game. Learning to be calm in the face of an enemy was an important lesson, as was learning when to run and when to freeze. It is much harder to stay calm when a person who has lost a democratic election is allowed to continue exploiting the very people who voted him out.

Originally, when I decided to include this story, I pictured the bees as a metaphor of government and its endless bureaucratic minions harassing the people. The more I thought about it, especially in the context of ousting bad politicians, I realized that we, the people, are the bees. We have already built our nests, and we swarm the election to rid ourselves of the office holders who throw rocks at our way of life. The lame duck politicians are the challengers, because they want to come as close to the edge of the law as possible and do as much as they can before they are removed from office. They know that we can sting, but they are willing to test us to the limit of our endurance. I leave it to you to research how the honey, i.e. campaign donations, flows to lame duck candidates during the interim between losing the election and leaving office.

I would like to see the law changed so that any legislative body affected by an election would be adjourned on the last Friday before voting day. (Those who draft the law could compose an exclusion covering nuclear war or other disasters that Congress would like to think it can manage.) As each election is certified, the winning candidate is sworn in immediately, and as soon as a quorum of elected representatives has been sworn in, the legislative body is allowed to reconvene and commence its work—if the phrase "commence its work" can be at all meaningful when referring to a legislative body.

> Item 5: "Citizens of a free republic have the right to demand that their electoral decisions be honored immediately, and as such citizens, we demand that each elected candidate be sworn into office no later than twenty-four (24) hours after the Secretary of State certifies the results of that candidate's elective race."

It is important that our wishes be honored by our leaders. When big money sways the political agenda in the direction of special interests, no politician deserves our undivided loyalty. Most politicians are for sale, and they deserve to be beaten down and out when they've proven that their loyalties have drifted from their constituents. We should never let someone that no longer represents the people cast *any* vote in OUR name.

"Now Absalom happened to meet David's men. He was riding his mule, and as the mule went under the thick branches of a large oak, Absalom's hair got caught in the tree. He was left hanging in midair, while the mule he was riding kept on going."

2 Samuel 17:9

Church of Christ, Coolidge, Texas demolished

IRS Form 1099-GOV

IRS Form 1099-GOV is an idea that I have been championing for years. It was the cornerstone of my pseudo write-in campaign against Senator John Cornyn in 2008. Like the schemes of progressive, social-warrior leftists, my idea revolves around "fairness," but with one big difference: 1099-GOV centers on fairness for all American citizens rather than insulated special interest groups. The government belongs to *all of us*, and each of us deserves to be treated equally before the law.

Does that make you laugh, too? After decades of watching politicians cater to groups with government largesse, it is hard to believe that we can find a way to turn that ship, but I cling to the hope that people who have been bought with freedom will remain loyal to freedom, unlike politicians.

"That is why we labor and strive, because we have put our hope in the living God, who is the Savior of all people, and especially of those who believe."

1 Timothy 4:10

My elevator speech for the Internal Revenue Service 1099-GOV went something like the following paragraph. Depending on the audience, it usually took about a minute. Sometimes I can talk fast.

Working Americans' incomes are tracked in every way. If you are a wage slave, you get a W-2 which reports your wages to the government. If you are an independent contractor or investor, you get a 1099-MISC. If you used to work and are now on unemployment, you might get a 1099-G. If for any reason you convince a bank to reduce the principal on your loan—because, for example, you claim you were suckered into buying it by an evil bank that allowed you to sign for it like an adult—that reduction in principal is income to you and the bank is required to file a 1099-C. When a friend of mine was the victim of $8,000 dollars' worth of credit card fraud, the bank investigated and wiped his account clean, but it also cut a 1099-C. The IRS went after him for taxes on the fraud. Really. It took months for them to clear his account. (Ed: It takes months for the IRS to do *anything* which doesn't involve destroying subpoenaed

```
records.)
   Contrast that with welfare, AFDC,
Pell grants, section 8 housing, free
transportation tokens, free
identification, Medicaid, free
electrical support for AC in summer,
heating oil support in winter, and any
number of other government programs
that supply actual dollars or services
to people. The government funds these
programs with our tax dollars, which
have become 'income redistributed.'
Why aren't the recipients of these
programs required to report these
benefits as income?
```

At this point, people listening to my speech were either interested or already walking off. During my campaign against Senator Cornyn, which was doomed from the start (Cornyn: $10 million, SenatorMark4: new hat and spurs), I never saw anyone turned off by the idea. I talked to crowds in Austin bars (lots), hung around Crawford, Texas to watch Code Pink harass the Bush Ranch, and I met Bush's entire Secret Service detail. I talked with Senator Cornyn at the Texas state convention about my 1099-GOV idea, and even he said it was outside the box. At the Senator's suggestion, someone in his entourage gave me his tax expert's credit card, and then they all disappeared. If you don't believe that an email to Senator Cornyn's tax expert resulted in anything, you'd be right.

The last I heard, anyone who pays an individual over $600 in one year is required to file a 1099-MISC, the form which reports those payments to the IRS. I know for a fact that plenty of government payments exceed that amount in one year. There was a time, I'm told, when San Francisco gave homeless people $300 per month until they couldn't afford to pay the ensuing flood of immigrant street sleepers. If responsible citizens are required to report every penny of their income, why shouldn't people receiving government handouts be held to the same regulations? If any one of us fails to file a 1099-MISC on our contractors, we are liable for at least $250 per statement with no maximum. Fairness? Only if the government *of* the people follows the same rules that it imposes *on* the people.

Sometimes with an IRS form, eventually a gun and a seizure of property, government lays claim to our money. They then pile it up, pass their hands over it a few times, and thereby transform it into their money. From that point onward, income reporting requirements no longer apply, because it is their money, and they'll do with it as they wish. They utilize it as "income, redistributed." Carefully. Efficiently. Yeah, right. Egregious cases that fly in the face of true fairness are too numerous to count.

When I finished my training as a hospital corpsman at Balboa Regional Medical Center in San Diego, CA, I was sent to Bethesda, Maryland. While awaiting the start of the Medical Technology, B.S. program, I was to work in the hospital wards. Working while waiting for school was called "being stashed," and before being stashed on the wards, I had to go through hospital in-service training to learn how to work in that world. The medical tech classes were weeks off, too, so I was stashed in an office staffed by civil servants doing "receipt control."

My perception now is that receipt control departments do nothing more than shuffle bills to make it look like they're actually tracking the money that they spill on the ground. I was assigned to assist a woman—let's call her Mable—because she was so far behind. Mable's desk was the only one in the office that was buried under files. I am not joking when I say that the smallest pile on her desk was over a foot tall. The paper forest growing out of her desk left little space for any office work except taking phone calls. It didn't take her long at all to train me, because she had a printed list of standard operating procedures.

The files crashing in around Mable were

reimbursement claims sent by people who had attended training at remote sites, and the claims included receipts from hotels, rental cars, and miscellaneous expenses listed on the claim forms as eligible for reimbursement if improved. Each file included about twelve pages, and each page required a specific number of copies for further processing by the bureaucracy. Mable's desk was the pit stop for filling the files with the proper number of copies, and her SOP was simple: Form 1: 6 copies; Form 2: 4 copies; Form 3: 1 copy; Form 4: 7 copies—or something equally diverse. No way could it be done in an automated fashion without wasting lots of paper, which was important enough to be included in guidelines: "No waste! Our budget only has so much room for toner and paper."

The job was straightforward. Pull the first page off, make 6 copies, turn them face down on the file. Pull the second …on and on and on, all day long. It was mind numbing, which explained Mable's countenance. After a few score files, the check list was burned into my mind, and as long I focused, things went smoothly. *Sheee...ee...eeee, chukbunk. sluuup.* Times ten thousand. Per day. It was the kind of job that furnishes sound tracks for unbidden dreams.

Fortunately for Mable, HM3 Robbins was totally dedicated to doing whatever was put in front of him to the best of his ability. I slashed my way through the files, neatly stacking the finished work on a cart which was delivered daily to some other paper maw in the Bethesda tower. By the end of my two weeks I was thoroughly proud of myself. Mable's desk was clean, but she didn't seem particularly thankful for the help. As the grand finale of my stay there, I announced that I was going to wheel the cart to the next stop, because I was curious about what came next. I was confused when they were reluctant to let me do more; it was as if I'd threatened a union job. They weren't happy about my delivering the cart, but I wasn't going to take a no for an answer, so off I went to the final task of my temporary stash.

At the time I was still fairly young and ignorant of the ways of the world—of the government's ways especially—and I was totally unprepared for what happened next. I delivered the cart, and stayed for a moment while the cute young receptionist thanked me for the cart. Then she shared her frustration with me. Her first job was to go through every file and rip out every copy except the original. All those copies I'd so carefully made like a trained zombie went straight to the recycle bag.

Every single copy.

In those days my blood boiled when obvious stupidity jumped in front of me. I don't remember my discussion with the receipt control director at all, but I have a vague memory of learning that Mable only had to make it one more year before she could retire. The thought that civil workers would continue to waste resources—and, being generous, an intelligent mind—for no purpose beyond insuring that their friend made it to retirement, still makes the hair on the back on my neck stand up. Think about how many people were aware of this and let it continue. Mable's coworkers surely knew that she was already "retired-on-active-duty." On top of that, she put on a continual whining act about too much work! As a young man I let it slide despite my disapproval, but today I wouldn't stand for it.

When I worked in Texas state government, I heard stories about "ghost employees," workers who received paychecks but never showed up in the office. I never encountered that kind of ghost, but all the old timers had stories. In her own way, Mable was a ghost employee, and there was no time during my government employment that I didn't see some of her type. I didn't have to look hard, either. Given the expansion of government, I grew to expect a concomitant growth in Mable's type of ghost job.

Ghost jobs aggravate me, but at least people like Mable have to show up for work to collect their paychecks. They have to be respectful of their bosses, and they have to fight the commute, just like the rest of us. What *really* bothers me is that our leaders feel that people who don't even pretend to work deserve some of our personal earnings. A case in point: some Democrat lawmakers in Colorado recently blocked a move to prevent EBT cards from being used in ATM's located in marijuana dispensaries. This kind of thinking is exactly why the 1099-GOV should be implemented. Recipients of government handouts would then be responsible for reporting their "income, redistributed" whether they paid taxes or not, and they would be required to do what regulation and law requires from the rest of us. If they want to get high on my dime, they should file their IRS forms, too. They want to share the wealth? Then they should share the pain, especially if they're going to shine the gain!

Baby boomers and their children know that government is out of hand, but convincing the millennial generation is tougher. Millennials don't really understand where government funding comes from, and they want to cash some of those checks, too. They are a generation very different from us on many levels. The thought of having a pointless job that allows them to collect money for sitting and texting all day is alluring to them. Most of them are not yet home owners, so they have no perspective on property taxes. Their sense of entitlement, however, makes them sensitive to having their income reported for the purpose of government gleaning. Something as simple as reporting the property tax share on a rent bill—much like notating how much interest has been paid on a yearly credit card statement—would go a long way toward awakening them to how much money their votes are costing them. Exposing the invisible hand of the tax man is essential if we are to restrain it.

"From whom do the kings of the earth collect duty and taxes-from their own children or from others?"

Mathew 17:25

Glenn Reynolds has been beating on the idea

that we should revoke the Hollywood Tax Breaks. I agree with him. What thinking person wouldn't view them as nothing more than a crony capitalist abuse of our system? It is shameful that government funds media adventures with our tax dollars, yet will not fund conservative ventures. Whose children are we feeding? Liberal media from California comes to Texas promising jobs, and in turn, Texas offers to pay for part of their propaganda. It doesn't take long to find the Texas, Austin, or Dallas Film Commissions; state and city employees are being paid to encourage left-wing companies to build left-wing media using our money. The follow-on question, obviously, is "How much?" If a 1099-GOV were required, and if the companies had to note it on their tax returns, we would know *exactly* how much. I'm still betting that demanding copyright for using our money would open those notoriously corrupt companies to a little Texas sunshine. Sounds like a place for a lawyerly attack to me.

Another way to convince millennials that our government is out of control might be a different approach to income tax withholding. As we all know, government takes a portion of our income before we see it, and the amount appears as a small line item on our reduced checks. The amount may seem a little large, but we justify it by believing that we all do it together. What if withholding was deposited in a bank and wage earners were required to sign it over to the Feds every quarter? I'd bet that when millennials started signing those checks, they'd be on high alert!

My conservative friends and I feel continual frustration when government acts like an employer, because "acting" is exactly what it is doing. Elected officials and bureaucrats are playing with money as if its only value is to buy votes. Double-entry accounting? It's totally Greek and rushing toward the same catastrophic outcome. They aren't even demanding personal responsibility from those who feel free to have the government buy their food and their pot. (Did I really say that? Have we arrived? Can we get off now?)

The revulsion I feel about the way government implements its policies butts against others who argue that government only wastes a small percentage of its revenue. They maintain that a billion-dollar rounding error of the people's money is a small price to pay for the welfare of society. Conservative voters have reached the point at which their revulsion stands on the edge of abdication. Government waste has become so common that nobody in leadership is able see it, either. Too many of us have become immune to the destructive policies of the government agencies which we originally installed to support the goals of our free society.

The mention of revulsion stirs up another Navy story. During Hospital Corpsman School in San Diego, I was the guinea pig. I didn't mind, in most cases, letting fellow students use me to improve their blood drawing skills. I allowed new corpsmen to practice inserting IV's in my arms and the backs of my hands before they did it to patients. I was known for it.

Some hospital wards require a certain "iron" from staff; otherwise, making it through the day is tough. At Balboa, as in most hospitals, one such place is the neurosurgical ward, and I volunteered to do my last days of training there. It was a very scary place. About thirty beds in an open ward were filled with men who had suffered all kinds of injuries. I have mentioned before that most of them were young naval personnel who hadn't learned the rules about diving in shallow water or riding a motorcycle without a helmet.

Three of us, two men and a woman, volunteered to serve our exit rotations on this busy and terrifying ward. Despite being warned about the atmosphere, we wanted to be brave and gung-ho, but there are no words to describe what our first few steps into that ward were like. Some patients had huge, grotesque balloons of scalp on the sides of their heads where surgeons had opened them to relieve pressure building inside their skulls. Most had received tracheotomies and breathed through ventilators. The sounds and sights and smells of damaged young men would make anyone gasp. The smells, though, made us want to hold our breath to avoid inhaling the thick, almost viscous air that smelled of "not well." The nurse, a Navy Lieutenant, met us a few steps inside the door as we stood there frozen. She introduced herself, and then, all business, she led us to a seriously injured patient to get us started.

She presented us to a patient who had one of those balloons of fluid on his head and a trach in his neck. She explained how to change the sheets and the Foley bags, and she explained other procedures that were already familiar to us. Then it was time to learn how to suction a trach: we were to disconnect the ventilator, drop a tube down the patient's trachea, cover a hole in the tube which is attached to a vacuum, and suck the phlegm out of the patient. The procedure would make the patient cough, which would bring up more. (Being static and immobile makes a person's lungs fill with bad stuff.)

I was standing at the nurse's right elbow, a bit removed so she would have room to work. My colleagues were on the other side of the bed. Like all patient care professionals, the nurse carefully explained to the patient what she was doing, and we were listening, too. She removed the ventilator connection, cleaned the trach, dropped the tube down his lungs, and the horrible slurping sound as she applied the vacuum and removed the tube was almost too much. We all caught each others' eyes at this point, which was a microsecond before the next event. In an instant, the patient's whole body wretched, and a record-breaking loogie exploded from the trach tube. It grew and expanded, covering the nurse's head with a perfect sheet of phlegm. Completely—her face, her hair, and down to her shoulders. A plastic shopping bag tied around her neck wouldn't have covered any more. Oh.My.Lord!

"...and provide for those who grieve in Zion— to bestow on them a crown of beauty instead of

ashes, the oil of joy instead of mourning, and a garment of praise instead of a spirit of despair. They will be called oaks of righteousness, a planting of the Lord for the display of his splendor."

Isaiah 61:3

We three, fresh Hospital Corpsmen locked eyes. In less than a second, the female among us decorated the patient with projectile vomit and then ran out of the ward, crying. The remaining trainee looked at me and then started walking to the door, but he didn't make it, either. He decorated the floor twice before he ran out, too. I turned and looked at the nurse as she calmly, almost robotically, located some wipes and started cleaning herself up.

"Are you OK?" she asked. To this very day, I can see her face, dripping phlegm as she spoke. It was a horror show.

"Yes, but I think I'm going to go smoke a

cigarette at the elevator," I said. I don't remember my voice sounding strained, but surely it must have. I do faintly recall that my walking to the elevator was a little wooden, and it seemed that all eyes were watching me. I have since wondered if new volunteers on that ward provided an ongoing office pool.

This level of revulsion is exactly what free people should experience when they discover their government pocketing tax money or pouring it down rat holes. Immediate and writhing revulsion is the only proper reaction, but that's not what we're seeing, is it? Most of us have become inured to it, watching and expecting someone else to take action because we're too weary to take it on ourselves.

"When the Israelites were also mustered and given provisions, they marched out to meet them. The Israelites camped opposite them like two small flocks of goats, while the Arameans covered the countryside."

1 Kings 20:27

This is why I think the 1099-GOV would be a perfect way to regain some control over our government. The form wouldn't be entirely new; it would be like a 1099-MISC except for "income, redistributed." We don't need the Feds' permission to file them. In fact, if we simply issued them as "1099-MISC (GOV)" from state government agencies, would they really just throw them away? Maybe that's not a good question, because we know that they're capable of almost anything, but if we filed them as 1099-MISC, what could they do but put them into their database? Let's not forget that their database is the tool that makes them think they control us.

The best part of the 1099-GOV idea is that the technology to handle them is already in place. *Every single government agency*, at every level, already issues W-2's to all their employees and 1099-MISC's to all the contractors who work for them. Though it sounds impossible, I once bragged that, given the data library for the existing 1099-MISC reporting system and the database for all checks used to pay "income, redistributed," I could port the data for the first beta report to the 1099-MISC system in two weeks all by myself. I am older now and out of practice, so maybe three weeks. I'd bet that it would be easy to locate plenty of hungry bitheads who could put out the first beta report in the space of a long weekend, because the data port is a one-to-one relationship. If the Texas governor were to demand such a report from the payments system vendors, the vendors would produce it —without cost—to maintain the business relationship.

The idea of a phone call resolving a fairness issue is tempting, is it not?

What intrigues me about the 1099-GOV idea is that it puts a fence around what government can claim is theirs. When payments are made to individuals regardless of their actions, the total amount of those payments defines a baseline tax deduction for responsible, working citizens. If someone receives $300 per month from the government for no other reason than breathing, then every American citizen who breathes *and works* must receive at least that same consideration as a minimum. Using the same $300 figure, this would mean that a family of five could claim a base deduction of at least $1500 per month on this consideration alone.

But equal compensation for everyone who lives and breathes isn't the "fairness" that we're looking for, is it? We all know that a person who showers, fights the commute, and smiles at the boss is not like someone whose only labor is cashing an EBT at a pot dispensary. The laborer deserves something more. The idea that a working man or woman deserves the same consideration as a state dependent does not strike me as outrageous. It doesn't strike me as fair, either.

"Don't I have the right to do what I want with my own money? Or are you envious because I am generous?"

Matthew 20:15

When I started playing with this idea in earnest, it hit me: Americans could use a mathematical equation to describe a fair social support system. The only premise required for accepting the equation is the belief that citizens who work for a living deserve as much consideration as those who don't. Translated, this means that people who slug out a living to house their families should be considered as important as those who smoke EBT-purchased marijuana in their section 8 homes.

I put my equation on a sign and took it to the Republican State Convention. I also marched it around Glenn Beck's studios in Las Colinas. Only the police were curious about my badge, and in the space of fifteen minutes I was introduced to the duty staff of the Las Colinas Police Department. I think I even met the chief, too, and I learned that if there is no sidewalk beside a building, then there is no public easement, either. The department liked my idea, they said, but they let me know that a sign intended to give a math lesson to motorists was a little much. They loved my spurs. But...move along, please.

```
The equation is simple: 1099-GOV =
FAIRNESS x (LABOR/2)
```

1099-GOV represents the amount of government assistance, i.e. "income redistributed."

LABOR represents the incomes of working Americans, *before taxes*. Thinking generously, I decided that 50% was the highest percentage that American wage earners would *ever* allow the government to take from their incomes, so that is LABOR divided by 2 (LABOR/2).

Therefore, (LABOR/2) is the worker's TAKE-HOME pay after socialists massage it.

If we let FAIRNESS = 1, i.e. unity and perfect equality, the equation says that those who live on government assistance are entitled to receive the same amount taken home by people who work. I consider that unreasonable, and I hope you do, too. So how do we determine fairness? Are people who don't work entitled to 10% of a working citizen's income? 50%? Although 50% seems outrageous to me, let's keep things simple (and generous, for our leftist friends) and use it. (It's socialism all the way down now, isn't it?) Therefore, FAIRNESS=0.5 for this example.

Manipulating the equation mathematically

makes things interesting. Dividing both sides by FAIRNESS, we arrive at 1099-GOV/FAIRNESS = TAKE-HOME. OR (1099-GOV)/ 0.5 = TAKE-HOME

Using the previously mentioned $300 figure given monthly to San Francisco street sleepers, a working person's take home, tax exempt, should be $600. Therefore, a family of five should be able to take home, tax exempt, $3,000.00 per month before the IRS is interested in them. That's $36,000 for a base exemption for being a working American. That doesn't even include all the other "income, redistributed" that flows from government.

Such an equation could be a valuable tool to limit government spending. Whenever the government decided that it wanted to spend more money on entitlements, it would be real entitlements for the entire citizen base, and, real income from working people would be excluded from capture at the same increase that entitlements were growing. *That* is fair!

This stance on fairness is exactly the place from which a smart, courageous politician could attack the left in an effort to give the middle class *the same relief* that government gives to the indolent. There is no reason for government to categorize people according to color, religion, or even the place they choose to live. If a street dweller in San Fran gets $300 per month, why shouldn't a working man in Midland receive the same amount against his living expenses? If one gets a 1099-MISC, so should the other receive their 1099-GOV. Fairness. And do remember, the IRS doesn't really care if they make you a street sleeper for failing to follow their rules, so street sleepers would suffer the same criminal penalties for failing to file *their* returns.

Such an arrangement would, of course, lead to arguments about whether the median or the average should be used to calculate the 1099-GOV comparator, and I can certainly see that written regulations would be bounced around from every corner. Should a man working in the hot sun actually get more than a man that …? Take your pick. I can hear The Loud now. This would certainly lead to a new flood of regulations but at least they would be pertaining to the equal treatment of each American.

But regulations, like laws, are more often ignored than we care to admit. Recent Supreme Court rulings make the idea of "black letter law" seem quaint. We are increasingly a country ruled by men rather than by law, and we are reminded daily that those who live by a creed of honesty are fading from the public arena. There doesn't seem to be any rule in any area of life which hasn't been ignored or violated by people in elevated government positions, nor is there a regulation or specification that union workers can't ignore.

When I served in the U.S. Navy as a regular line officer, I was once the Main Propulsion Assistant on a frigate, the USS Talbot (FFG-4). At one point we went into a yard period at the Philadelphia Naval Shipyard to refit the ship. My men owned more than 180,000 man hours of that job, and the shipyard was responsible for even more. We were a small, single-screwed vessel, and too slow to keep up with the bigger ships, which maneuvered around us as we took the bee line. Frigates held very responsible positions in the carrier task forces: our purpose was defined by those aboard as "missile sponge." If an attacking force painted us with their fire control radar, our advanced technology allowed us to reflect back "aircraft carrier" (hit me, hit me, hit me). Please. Not.

A ship that cannot maneuver well, especially one that acts as a missile sponge, is "dead in the water" for real. My responsibilities were the main engineering spaces, and the captain had made it clear that **all** engineering watches would follow regulations *to the letter,* even though the Talbot was fourteen feet above the deck of the dry dock. Imagine how hard it was to get crewman to take soundings in the bilge water when the ship was high and in the sky. My solution to that problem was to put a note in a sounding tube cap before every watch: "Call Ensign Robbins RIGHT NOW," and I would know quickly whether or not the watchmen were being true to their task. It didn't take long for people to learn about me.

I worked very hard to maintain this same attention to detail in all the shipyard tasks. To say that I had an 8-feet stack of job requirements, all finely sorted in three ring binders, is no exaggeration. It was taller. Every listed job lined out all the tasks in detail.

One job that caused a memorable problem had to do with the main steam pipes. The Talbot was old and had been fitted with asbestos lagging on the 1200 psi main engineering piping systems. Once it was removed (a hermetically-sealed spacesuit job), the task was to "wire brush by hand to bare metal and apply high heat aluminum paint" to the steam pipes. The idea was that a grinder could weaken a piece of 1200 psi pipe and eventually cause a disaster. You only have to experience 950-degree steam leaking into a closed space once to understand that.

I could share numerous examples of the mid-watch shenanigans performed by union shipyard workers, but one in particular still warms my heart. One day, I was serving as Command Duty Officer, which meant that Ensign Robbins was the senior line officer and the master of all within our vessel sitting up on blocks. I was touring my engineering spaces and found a group of union painters using mops to apply high heat paint onto rusty steam pipes. I told them to stop and called their foreman. After some time, the foreman came down to see me, and I showed him the job description. He agreed and said it wouldn't happen again, but I made him initial and date the job description. The union workers didn't appreciate the foreman's yelling.

A couple of hours later I checked on the workers, and they were acting like painter pigs again. I called the foreman as soon as I saw it, and we walked into the space together. This time, I told him that if his crew didn't do the job as specified, we'd have to do something about it. He yelled again. He initialed the document again, this time under protest. I thought the painters might take a more professional approach after being told a second time because they were the brunt of more decibels, but an hour or so later, they were failing once again.

That was the limit. I spied them doing their lackadaisical job from the hatch to the main engineering space, called the Master at Arms and took him down to engineering with me. I told the union guys that I needed them to leave the ship.

"There is no friggin' way you're going to get us off this ship, ENSIGN Robbins!" they said. They were shouting at me, but I just nodded.

"I'm the CDO today, so I bet you're wrong," I said, puffing up a little. "You need to leave, or I'll have you removed," I said calmly. (Gosh, did I feel powerful saying that.) After some shouting and the application of handcuffs, they were marched off the ship and released onto the pier. I told the watch at the ship's prow not to let them return.

For the remaining hours of the night, I went over the incident in my mind a thousand times a minute. Putting shipyard union workers in handcuffs seemed like exactly the ticket. I really did enjoy it, but as the sun rose and the black sedans started lining up on the pier, I didn't feel so sure. In no time, I was informed that the unions were blocking the gates to the shipyard. I was relieved of duty and summoned to a meeting.

There I was the guest of honor at a table

lined on one side with full bird Captains and flag officers—all of them in perfectly pressed whites—and tired-looking union representatives on the other. I'm pretty sure it wasn't even 7 AM yet. The admiral sitting at the head of the table asked me to run through the incident, and I did, making sure to show the initialed job descriptions. When I was done, they thanked me, and I left. I took a deep breath at last. The union strike went away like vapor, and that was it. I never heard another word about it.

I personally caused a strike that impacted tens of thousands of people trying to get to their jobs. Although I doubt it lasted two hours, people were still lined up at the gates when I went to see what it looked like after I left the meeting.

I don't react quite so quickly now. On the other hand, I will never keep anyone on *my* staff who can't or won't perform a job properly. I'd like to believe we can elect conservative representatives who feel the same way. Maybe someone who actually had to make a payroll would understand this.

John Edwards, a disgraced Democrat who was indicted on charges of violating campaign contribution laws, always ran on about there being two Americas, the haves and the have nots. In my opinion there are three, and I can clearly define them. First is the professional man, doing so well lately, who showers in the morning to charge up for the day. Second is the working man, who showers at night to wash off the remains of the day, or his wife won't let him sit down to the table. Third is the America defined by those who don't shower at all. The Democrat Party seems to care only about the morning showers and the no showers. The 1099-GOV would protect the people who shower at night.

    Item 6: "Texans believe in self-
    reliance and demand that the
    government treat all income, earned or
    redistributed, with equal reporting
    requirements; toward that end and in
    order to spread fairness throughout
    our government, Texas introduces the
    1099-GOV."

If the 1099-GOV forced the income redistribution industry to report their income as society's working citizens do, we would hear a collective cry that would be beautiful to behold. It would be even louder than the shipyard union strike and would make my "measured" response to lazy workers seem patient. Treating all Americans with equal consideration is the kind of fairness that liberals can't tolerate, because they know that their power comes from defining special interest groups.

In reality, when leftists announce that they are going to work for social and economic justice, they are really saying that they intend to work at the edges of existing black letter law in the greatest democracy the world has ever known. You don't hear them mocking Cuba; instead they make movies extolling the virtues of the Cuban healthcare system, and they totally ignore Cuban citizens' lack of freedom. Elites from the left dine with dictators and excuse their failed economic justice experiments, all the while demanding that Americans supply their forgotten people's needs because we are successful. Making *our* government conform to the standards that they impose on us—the ones who are legally *their* masters—is the first step toward accountability and true economic justice, and would be transparent insurance against the assaults of Marxist justice warriors.

Venezuela is a perfect example of gangster government. Though officially a republic, Venezuela has at its helm President Nicolas Maduro, an authoritarian socialist. In an effort to solve his country's economic woes, Maduro has requested aid from Russia and China, and his government's agreements with China have only complicated Venezuela's failing economy. Food is scarce, and citizens line up for blocks to wait for basic goods. The government has taken over a major supermarket chain, and Maduro has arrested its owner on charges that he was hording food to hurt the country's economy. Maduro has arrested several entrepreneurs for protesting government policies. The major difference between what is brewing in Venezuela and what we are cooking here is the element of time. The leftists in our government are just a little behind because our Constitution blocks their way. In order to prevent despotic behavior in our own government, we must insure that the Constitution remains the primary rule of law, and that government officials be as accountable to it as all other Americans.

As a response to several corporate accounting scandals which have cost American investors billions of dollars, Congress passed the Sarbanes-Oxley Act, which outlined new and expanded requirements for all American business management and accounting. Its provisions include guidelines for full financial disclosure, third-party auditing and criminal penalties for misconduct. Interestingly, the provisions of this law *do not apply* to the

United States government. Imagine trying to hold public servants criminally responsible for financial misconduct. It can't be done until we demand the same accountability from government that we try to impose on corporations, so let's add:

> "In addition, no taxing entity may levy taxes without a biannual audit confirming their adherence to GAAP and Sarbanes-Oxley."

That line will cause all kinds of screaming. I can hear it now: "Oh, we're a cash business" (we don't need to track your money using standards). "We don't pay enough to hire the quality of personnel who are willing to take legal responsibility" (for your money, so we just spend it without standards, knowledgeable supervision, or accountability). I admit that this provision would add audit costs to government, thereby impacting services. It would be a point of contention, but I hope I'm not alone in believing that uncomfortable adjustment is better than the spillage of funds that we see on a daily basis. I believe my money deserves greater respect. When my funds go up in marijuana smoke via an EBT withdrawal, I'm not very sympathetic to this complaint. How about you?

It would be great to add a provision like that to the preamble, but because it goes against the entire government status quo, I look at it as a very steep hill to climb. Still, the idea that we Americans could starve government if it didn't properly account for *our* money is appealing on so many levels. It would be the ultimate reward, a perfect example of **operant conditioning** and effective management, and the loudest public punishment for those who refused to manage our money properly. That's why I look forward to seeing whether this provision is written into the preamble. The fact that Congress excluded itself from Obamacare reinforces the necessity for demanding accountability from government, and that demand absolutely must come from the grassroots.

"My bones suffer mortal agony as my foes taunt me, saying to me all day long, "Where is your God?"

Psalms 42:10

"YOUR" Government Accounts

In order for a people to defend freedom, they must know what their government is doing. Money is the blood that feeds any organization, but unlike commercial organizations, government is not required to balance its books. If we intend to regain control of government, we must focus our first efforts there, and in order to do that, we must gain control of the books and records themselves.

```
Your government records.
YOUR GOVERNMENT records.
YOUR GOVERNMENT RECORDS!
```

Even after saying those words out loud a few times, most of us don't believe them. I don't believe them. Nothing in our experience makes us truly believe that we have the right to understand what government does in our name. Even though the Freedom of Information Act was signed by President Johnson in 1966, our fifty years of experience with it doesn't soothe those of us who want to know what government is doing. And to repeat: Sarbanes-Oxley doesn't apply to government, and Congress excluded itself from Obamacare. Folks, we got issues with Uncle Sam's trustworthiness.

Our distrust has been validated by watching the IRS bob and weave around Congress's subpoenas for Lois Lerner's records. If elected Congressmen with subpoena powers can't investigate corruption or get answers from the government in a timely fashion, what hope do ordinary citizens have? Would any of us really want to chase the IRS and stir their ire? Although the Freedom of Information Act only applies to Federal agencies, every state in our union says that citizens should have the right to know what their government is doing. But, as in all things, talking the talk and walking the walk are two different things

In the mid 1990's I was a geek at the Texas Natural Resources Information System. TNRIS is a small agency that keeps track of a myriad of interesting things—political boundaries and water well logs, to name just two. This agency is wholly contained in the Texas Water Development Board. The internet was very new then, and I was responsible for all the UNIX systems in both agencies. I was also the one watching the protocol analyzer that snooped on the traffic entering the agencies' gateway.

The engineers loved to play with the computers and when the UNIX boxes got sideways, I'd spend significant time sorting out the problem. The significance of a comma in a shell script is not to be denied, so in order to fix the problems, I found and used a book called **"Using UNIX to Audit UNIX"**.(I don't remember the author's name. The book is out of print, and I haven't been able to find it on Google.) So I had this book that told me how to audit UNIX systems using shell scripting. I implemented it on all the systems and then converted it to PERL so it was much faster. Thanks to the book, I was able to get things running smoothly and diagnose problems sooner, and I also learned to appreciate the power that the audits gave me.

UNIX systems can create large amounts of logging information, which makes scanning log files too difficult to manage efficiently without automation. In order to prevent my having to look over dozens of audit logs, I had implemented a scanner that extracted the important log entries and sent them to me in an email every morning. The list of entries that the scanner looked for was growing all the time as I learned what to monitor, and one morning I noticed a weird log entry.

One of the things I decided to watch pretty early in this evolution was late night file creation. As the normal files were slowly eliminated from my alert email, which cleared my vision of the agency-wide activity, I came across quite a bit of activity coming across to one of the TWDB hosts. It was all late, and it all came over as "root," which meant that a superuser was logging in remotely and running stuff on one of the main hosts used by the TWDB to do their geographic information system (GIS) plotting. *NOT good!*

I knew that the Texas Water Development Board had nothing at all to do with oil and gas. Their job was to help small communities develop water and waste treatment facilities by buying their development bonds. As my investigation into the interesting logins proceeded, I found some maps that were being developed for a gas pipeline in Chile. South America. I knew that the citizens of Texas would not appreciate their resources being used for that, so I decided I would slowly—one management level at a time—walk it up the chain of command. Calmly. I believed I knew what to expect.

I printed out the GIS map and started the long march up the chain of command, pausing two weeks between each step. To paraphrase the different responses:

```
Level 1: "Thank you for your
attention. I'll look into it."
Level 2: "We heard that and haven't
made a determination."
Level 3: "Interesting. Thanks. We'll
```

```
get back to you when we know more."
   Level 4: "What's the big deal? We
don't use them at night anyway." I
have to admit that this one stunned
me, coming from the second in command.
```

I proffered that the Texas Department of Transportation had bulldozers and graders that they didn't usually use at night either, and my family had some ranch property. "Do you think it would be alright if I took their equipment and just made sure the fuel was replaced when I returned it in the morning?" I blurted out, just a little incredulous.

"That doesn't even make sense. What's your point?" he said. I just nodded, and we ended the meeting.

By now I'd reached the Executive Administrator level and it was obvious that some decisions had been made. I never put a name on what I thought was going on, but two people on staff were now allowed to work from home, which was unheard of in those days. I implemented a policy, with management approval, to block any "root" logins from the internet.

I decided that I'd write a certified letter to the Executive Administrator and copy the General Counsel, providing a short synopsis of the issue and requesting a meeting. Sometime later I received the return receipt from the EA's letter, but I never received one from the General Counsel. I thought that was interesting, so I asked for an appointment, and the conversation followed this outline closely

```
    "I sent you a certified letter, but
I didn't get the receipt back", I
said.
    "Well, the EA got your letter…," but
I interrupted her.
    "That's not why I'm here. I know he
got my letter, but I didn't get the
receipt back from the one I sent you,"
I repeated.
    She started again, "Once the EA got
your letter, he decided…," but I
interrupted again.
    "Oh, that's OK. I'll make an
appointment with him right away." I
finished and left.
```

I called the postal inspector to find out what happened to my certified letter. He informed me that at government agencies, they rip the receipt off in the basement, and they sometimes get lost. Not on purpose, of course. If you have something important to say, send it via **registered** mail, which records a chain of custody, which means that no one can claim not to have received it. This is very important when dealing with government.

I sent the postal inspector a *registered* letter

requesting information about my certified letter. Within days the general counsel found my original certified letter under a stack of stuff on her desk. She said. I'm sure it was just coincidence that she looked for it again just days after our meeting.

In the end, I had my meeting with the EA, we covered all the points, I told him I had things under control, and that was the end of it. We never named names or discussed legal implications. Looking back on it now, I wonder if it might have been our administration working with Chile to make South America more self-sufficient in natural gas. It follows right along with our making sure that Iran is self-sufficient in nuclear power. Government is just another name for those things that countries do together. In secret. Out of the public eye.

My experience taught me that any dealings with the business of government are likely to be covered up and buried if they cause the public to question them. Elected officials and bureaucrats really do think of themselves as the "business of government," and they don't believe that they can be replaced. They believe that they diligently do the work of the people, and that troublemakers deserve to be sidelined during their holy mission.

Would you be surprised if, in a post-911 world, those people were still working for Texas state government? You shouldn't be. When the Transportation Safety Administration hires people who were on its terrorist watch list, you know that anything is possible.

Making the government more transparent should be easy: just make their actions public. It's not naive. It's entirely possible if enough people insist on making it happen. If mandated transparency were a line item on a platform, for example, citizens would actually have a voice in determining it. An old-timer in Texas state government once told me, "Don't email anything you don't want on the front of the Houston Chronicle." I can promise you that our elected officials don't follow his advice. If they did, accessing records wouldn't be so hard. One of the key phrases in the Texas statute is "Cooperate with the governmental body's reasonable efforts to clarify the type or amount of information requested." What this means is that if you're on a fishing expedition, you can't expect them to drop everything they're doing to bait the hook for you. And they won't.

I'm sure that if government records became immediately available to the public, we'd have a more disciplined staff working in our state agencies. We would see better grammar and more direct pronouncements. The down side is that we'd also see more attempts to use non-government resources for back channel communications. Hillary isn't the first to maneuver outside the system's radar, and she won't be the last. Making all records immediately available prevent corrupt officials from hiding their use of government-supplied resources, which should be a selling point for public records being *truly* public from the start. Dealing with government is extremely difficult when officials are determined to slow roll information requested by citizens or Congressional subpoenas. Hardly anything in agency pastures gets more sustained attention than preventing potentially harmful information from being released. This is true at all levels of government boundaries: federal, state, and local. I know this for a fact.

I replaced the Texas Education Agency UNIX administrator in the late 1990's and immediately walked into a storm. After very little review, I realized that hundreds of the shell scripts which were running processes on the production systems were totally inadequate when viewed through a security lens. Imagine scripts on the boxes containing the hard-coded passwords for all kinds of computers and state education data. In fact, this was not just a state issue, the sign-on screen for the TEA main system made a point of stating that unauthorized access to those federally funded systems was a *federal felony*.

I noted the issues to my supervisor, and with some trepidation on her part, I went about finding and correcting the issues. I broke very little doing the corrections, but I discovered that someone who was very knowledgeable about the system configurations and user ID of the previous administrator was accessing the system. Did I mention that accessing federally funded computers which contained private education data was a federal felony? This was *surely* something that management would take seriously.

I talked to my supervisor. She thought I was trying to cause problems for her friend, but she'd tell him to stop, even though I didn't say it was him. Strange, that. I talked to her supervisor a few days later, who said that they were looking into it, and also that the old supervisor had been put on notice. I requested a meeting with his supervisor, and I was invited to a meeting around a table with about eight people. The general gist of the conversation was that they'd told him to stop, and he was a nice guy.

I filed a Freedom of Information Act letter

requesting certain logs from certain computers on certain dates. I didn't, like Congressmen, demand the world in a subpoena so that nothing would be destroyed. I knew what I wanted and knew exactly what it looked like, but I wanted the letter to pass through their hands so they couldn't deny knowledge of the unauthorized accesses. I can't remember if it was just before or just after this FOIA filing that I was put on administrative leave for 90 days. No change in pay or position, though. I just got to stay home and get paid, exactly like a ghost employee!

"Surely they intend to topple me from my lofty place; they take delight in lies. With their mouths they bless, but in their hearts they curse."

Psalms 62:4

I was obviously furloughed for exposing a federal felony, most likely perpetrated by someone they knew, but instead of being patted on the back for uncovering it, I was unofficially fired. Ninety days wasn't an arbitrary number, because if any agency fires, demotes or reduces the pay of an employee inside of ninety days, it must defend itself against charges of violating the whistleblower protection laws. When the agency fires the same employee at ninety-one days for whatever make-believe charge it has now inserted in its personnel jacket, the employee must assume an offensive position and prove his charges in court. This is the way that the scam of government transparency works. All the attorneys I talked to wanted money up front, because a fight with a gorilla that can stay in the ring for five years means that the process is expensive and the outcome is uncertain. I didn't want to pay for that, either. I decided that having a ninety-day paid vacation while I looked for other work wasn't all that bad.

Anyway, back to the requested paper. They called me and told me that the documents I'd requested were available for my review. If I preferred, I could have copies for ten cents or so per page, but they couldn't estimate how much it might total. I knew then that we were playing some kind of game, because the documents I'd requested couldn't have been five pages. I told them I just wanted to come in and review them. I had my laptop with me. A woman led me to a closet, as I remember it, and there on a desk were almost two feet of paper in individual sheets. I said something like, "Wow, it looks like you've been very thorough!" I could tell that she was happy with the surprise on my face, and I was working it, too. "Is it alright if I plug in here?" I asked, and she immediately nodded and left. I went to my car and retrieved my own personal copy machine, and when she saw me making copies on my own machine, the surprise jumped the space between us. I'm guessing that the agency didn't want to fight about letting me use their electrons for a copier after giving me permission to plug in.

How times have changed in a little over a decade! Today, I could see them announcing that I could review the documents but couldn't take in my smart phone. Doesn't "freedom of information" imply that we can leave the room with it? Actually share it in order to hold the government accountable for its actions?

At that time Texas had just become a red state after being a deeply Democrat state for almost 150 years. Austin was (and is) still the blue heart of Texas politics. I took all my information to the District Attorney's office, where they listened and reviewed my papers. I don't know whether or not the feds were notified, but I'd bet that a liaison was at least given a heads up about the felonious use of a state computer. In any case, that's all I ever heard about it, and I didn't expect any more because their smiles said it all. You don't fight gorillas.

Getting any organization with embedded interests to come clean about what they are doing will always be a problem, which is the reason I want to start with a structural change in government transparency. The gap between a government entity and the bank it uses is the place where accountability becomes difficult. Once funds enter the government mill, they are stamped and renamed and recycled from account to account with very little oversight. If the public could see all banking transactions *before* government accounts owned them, we might be able to regain control of the system.

The sheer volume of this transaction flow is an intrinsic problem. How could anyone make sense of it? After all, government itself can't make sense of it, which facilitates the type of misbehavior for which large organizations, especially government, are noted. When Congress asks the CIA for documents and they receive 7.3 million of them, we all know that the CIA isn't trying to be any more helpful than TEA was with me. They know that the Herculean task of reviewing all those documents will give them cover until they retire. The complexity of the process really does provide enough camouflage for evil to prosper.

Technology may provide a foil. How about crowd-sourcing all those documents? If they were available on the internet, and if 7.3 million people cared enough to look at one document and make notes which might catch another user's eye, all those documents could be initially reviewed in *one day*. If every comment added to a transaction changed the way the document looked on the screen—if, for example, the font changed, the font color changed, or the comment blinked—it would call other users' attention to an interesting item. By day two, there might be hundreds of issues calling for review, which is exactly what Congressional committees need. Instead of stopping the flow of information by burying us in paper, they would be required to answer questions asked by concerned citizens. I know that there are enough interested and caring Republicans in Texas to accomplish this sort of task in a week. *Doesn't it sound like fun?*

I can see three problems, here. The first is getting the information on the internet, but this is the easiest to solve, especially if we begin with government's bank statements. For one thing, bank deposits are the place where citizens' ownership line hasn't been blurred, and for another, banks already make account information available on the internet to millions of their depositors on a daily basis. I hesitate to say it, but doing the same with government accounts should be trivial. Passing a local ordinance stating that no bank can hold a government account without providing this feature would start that ball rolling without having to use notoriously inadequate government programmers.

The other two problems are more technical. The first is how to insure that the activities of 7.3 million people are coordinated efficiently. I have some ideas. When people enter the system, they could review and scan what's already there, or they could hit a button that says "crowd source." When the crowd source button is hit, it could deliver a small subset of the next transactions for review, and in this way, every person interested in crowd sourcing the data is sure not to do redundant work. If 7.3 million people log on and hit the button, then all 7.3 million documents would be reviewed at the same time, but a well-planned system should be able to handle heavy internet traffic. Multi-user online games host tens to hundreds of thousands of players simultaneously, and that internet world is far more demanding than one that would display a page of text.

Of course, there are some people who wouldn't want this to succeed, and they would enjoy hitting "crowd source" continually without making notes or comments thinking that they were denying others the ability to source the documents. Their efforts would have no effect on a system that rolled the pointer to the next crowd source document through the database, over and over again, so that any user could participate regardless of people trying to deny service. There would be some administration issues if, say, the CIA were to select a few hundred users hitting "crowd source" every few seconds in order to dominate the system in order to hide access to their documents, but I don't see city or state governments getting too involved in this kind of misappropriation of a system. They already have enough to do. Besides, any new system installation requires tweaking and refining. A simple approach to this problem would be to set things up so that any single user could only crowd source a set every two hours. That would put sand in the gears of anyone interested in causing problems. And let's be serious: eventually, we *will* find 7.3 million real Americans interested in sorting this out.. in one day!

The third problem is figuring out how to make it easy to enter the system and see what has been reviewed. I mentioned font size, font color, and possibly blinking fonts to make it easy to scroll through the system and see what people have been interested in documenting. Those are all great, but in this system we also need the ability to hit a button labeled "Hot Items" to identify any group of transactions within a tranche of documents that have drawn the attention of investigators, had lots of comments or critical comments, and thus have been flagged as a "Hot Item" by the system administrators. This option would make it possible for any committee to see how documents are reviewed and analyzed in real-time.

In the appendix I have attached a set of MySQL data structures which do all of this and more. Once implemented in any area, its major limiting factor is scanning/uploading documents into the system. I know that this would be a problem, because when people of every political stripe aren't outraged by someone like Hillary, who supplies paper copies (tree residue) of subpoenaed emails (electronic bits), legislation addressing the difference between the media is the only answer. I trust, maybe naively, that government will eventually address it. We may have time.

```
    Item 7: "We, the people of Texas,
deserve to know the workings of our
government at all times and demand
that all state bank account records be
made public and immediately available
on the internet."
```

I can already hear the shrill cries from agency heads about the need to protect the "people's" information. If I haven't convinced you that they don't mean it, at least look at the millions of records the feds have turned over to the Chinese in the Office of Personnel Management fiasco.

"For now we see only a reflection as in a mirror, then we shall see face to face. Now I know in part; then shall know fully, even as I am fully known."

1 Corinthians 13:12

United States Sovereignty

Sovereignty. I don't even know what it means anymore. When I was younger, my understanding of the term followed right along with today's Wikipedia: "…the full right and power of a governing body to govern itself without any interference from outside sources or bodies." The short description of how I view sovereignty is similar to the way I view my own home: if I caught someone snooping around the windows of my house, I would be free to remove them from my property, and if someone bashed in my front door, I would expect the law of the land to vindicate the violation of my legal boundaries.

I used to believe that a small collection of documents defined what the government was allowed to do inside and outside my home, and that the government's duty was to protect my rights as a citizen. I'm finding that belief harder and harder to maintain. Borders are no longer respected, and anyone can cross them. As a former computer guy, I know what it is like to have all the American citizens in a shop replaced by H1B visa holders. You don't have to live in Texas two days before you end up standing behind someone at a grocery store who can't communicate with the checkout staff as they offer their Lone Star (EBT) card. Anyone, illegal immigrants included, can claim welfare, and judging from what the government doles out, they deserve benefits even if they have lived in our society for less than a month. In addition, our leaders are sending billions of dollars to regimes whose people shout "Death to America" before their weekly call to prayer.

Democracy, welfare state, open borders.*Pick two.* I looked, but I couldn't find the place where I first saw this enhancement of Milton Friedman's postulation that no country could experience all three. It is the perfect description of what we are experiencing now.

I have never worked for a boss who wanted more democracy in the workplace, and politicians are exactly the same. Right now they are calculating what kind of welfare state they can promise to the world once they get rid of that pesky democracy. Obama is straining against the ropes of the Constitution to make his stance on this perfectly plain.

Tony Blair is reported to have said that we've entered a new post-Westphalian era because...globalization. He might have been right, because the English are currently lab rats testing his hypothesis: on the internet are videos of Muslims in the streets of London—in England, the first country ever to demand rights for individuals—waving signs that say "To Hell with Freedom." From my perspective, other ideals should be globalized, namely the First Amendment, the Second Amendment, and the right to legitimize government with free elections. Why would anyone in the Western world want to support what was defeated hundreds of years ago? If people want to reject freedom, can't we just take up a collection to send them back where they came from? Just because masses of people move to the United States for a better life, more welfare, and actual housing, it doesn't mean that American citizens should give up their hard-won freedoms. I hate to be a broken record, but instead of apologizing to other countries for being successful, we ought to concentrate on *spreading*freedom.

Back in 2000, I was driving down one of the streets of my childhood neighborhood in San Antonio. The streets were damp from a recent rain, and as I approached a stop sign, my anti-skid brakes decided to engage. I was barely moving, but I was drifting into a car at the stop sign. It looked like I might stop in time, but rather than pumping the brakes, I tried to push the peddle through the floor. As the distance closed, her brake lights went off, and I thought she might move out of my way. No such luck.

The picture shows that the 5 mph bumper was undamaged. You'll have to take my word that the scratch didn't take any primer off the bumper, even though it did leave a mark. No wrinkles. Fortunately, I had the good sense to call my wife and have her bring a camera to take pictures.

The police and an ambulance were called. Quite the flashing production. It was some time, maybe a year or two, before I received the first letter from a lawyer telling me that she was going to bring a lawsuit for $250,000 damages. I'm not saying that her foreign ancestry had anything to do with that; after all, I did scratch her car.

I sent them the picture. That attorney petitioned the court to be relieved because he was "too busy" to give the lawsuit the attention it deserved. It wasn't long before I got a second letter from another attorney. Sent the picture again. Same outcome. The third attorney was the same. The fourth attorney sent me a letter introducing himself and an 80-page interrogatory. My insurance company told me that I should hire my own attorney because she'd demanded more than my limits of coverage. I told them that if their lawyer couldn't get me out of this, then we were all doomed.

It took seven (7) years before we finally settled for $5,000. Granted, it happened before tort reform legislation was passed in Texas, but what a waste of all that time, scores of letters and countless meetings for something approaching $3 per day. It hardly seems worth it, but *someone* saw a payoff. Everybody knows "dollar."

We have truly entered a dark time. Our world is now a place where the great religion of peace, Islam, having finally been free from slavery for all of 50 years, is once again the center of slave trading. And yet…silence. Daily we read reports of the Islamic State executing human beings in the most grotesque ways, yet our United Nations leaders don't stand up to protest. The outrage over a killing a lion for $50,000 festers on the internet longer than the anger spawned by burning human beings alive. Americans make headlines by defacing statues of dead men who owned slaves over a century ago, but living slave traders rarely make the front section of any newspaper. Real, live slave market pricing for infidel women is published on the internet, and the highest prices are for children under ten. Ten!

We all seem paralyzed and unable to act. It is depressing in the extreme.

"God governs in the affairs of men. And if a sparrow cannot fall to the ground without his notice, is it possible that an empire can rise without His aid?"

Benjamin Franklin

There are 57 countries in the Organization of Islamic Cooperation (OIC), and from what I can gather, all of them condone the return of slavery. Arab slave trade is not just a bad idea; it's a stark, video-taped reality, which leads me to ask why our leftist media has remained silent about it. Like the pixelization that our press uses to keep us from fully viewing the videos of jihadist barbarism, the threat to our freedoms is purposely being veiled. The Prince of Lies has dropped technological wool over our eyes, and it's getting harder and harder to shake off.

Prime Directive

In the first season of the original Star Trek, the writers of the series introduced the Prime Directive, which was the guiding principle of the starship Enterprise's intergalactic travel:

> "No identification of self or mission. No interference with the social development of said planet. No references to space or the fact that there are other worlds or civilizations."
> Star Trek, 1968, "Bread and Circuses"

The Prime Directive seemed like fiction at the time, but a half century later, it is the mantra of our real-life leadership. It has had such devastating effects on the world's safety that I wonder if it could be anything less than satanic intervention. I can't imagine the devil implementing anything more powerful to serve his purposes in the minds of men. When the free world recognizes the "rights" of other countries to make their own mistakes—mistakes like starving, beheading and enslaving human beings—freedom is undermined. As an advanced people, what do we owe the world when we witness atrocities? What did we owe the world when Hitler marched through Europe? I read the other day that Islamic State terrorists were strapping children across their windshields as they drove into battle. Do we owe those children anything at all?

There was a time when we weren't so willing to accept the Prime Directive. Culture was analyzed and criticized on the basis of an objectively defined standard of right and wrong, i.e. of what was perceived as "good." Domestic battles were fought over women's rights and the abolition of slavery, both of which were advocated on the basis of that standard. The widespread belief that people are created in the image of God spawned the attitude that God expects free men to save people from cruel and vicious societies. If Cortez had been a Star Fleet officer, we'd still have human sacrifice.

Oh, wait. The Islamic State ties children to windshields as they enter battle!

When I was a kid, my friends and I loved to play in the sand pile in my back yard. It was shaded by two oak trees and was bounded by low stone walls, and every few years when the sand seemed to evaporate, my father would get a new load of sand and refill it for us. On a day after one of those refills, a friend and I were working the new sand. As little children, we had moved the sand around with toy trucks and bulldozers, but at age ten, it was engineering that fascinated us.

We had decided that we would dig from opposite sides of the sand pile until we had built a tunnel big enough to crawl through. Our plan was to meet in the middle. Being the engineers that we were, we dug our tunnels toward each other, but when they connected, they were slightly "off." When I tried to crawl through it the first time, I had to turn on my side to bend around the corner.

"While people are saying, "Peace and safety," destruction will come on them suddenly, as labor pains

on a pregnant woman, and they will not escape."

1 Thessalonians 3:5

Plop! The first sand dropped onto me. Shhhhhhhheeeeeeump! The tunnel caved in on me. I'm not sure if the air was driven out of the tunnel or out of my lungs, but I certainly heard a sound. Every part of my body was so closed in by sand that I was barely able to scream for help before my last breath was crushed out of me. For a few seconds I could hear my friend's hysterical laughter, muffled through the sand. Then he realized that it was no joke, and it seemed as though it took him forever to dig me out. Washing the sand from my nose was tough, the alternative was much worse.

Slavery is sand in the nostrils of freedom, and if we are to rid our future of it, we must act today. Our disdain for the merchants of human beings must be thorough, complete, and all-encompassing, and the evil which drives slavers must be extinguished. They may be primitive, they may be ignorant, and they may be a product of an antique culture, but their atrocities are an affront to decency, and we cannot allow their values to gain a foothold in our society. There is no place for slavery in our world, and if tenets of Islam justify slavery, Muslims must be forced to recant the justification before living in the United States, exactly as the Democrat South was forced to abolish slavery after the Civil War.

I intend to ignore the Prime Directive.

When WE Republicans marched through the south during the American Civil War, we burned Atlanta to the ground. Atlanta wasn't sacrificed because it was full of uniformed Confederates. (It might have been burned because it was full of Democrats, but....) The destruction of Atlanta was a declaration that supporting slavery in any fashion—whether by arms, hominy, or poem—was going to be punished severely. Citizens who disagreed were required to recant. Slavery was declared to be an affront to the goodness of America, to our Constitution, and to our God. It was then, and it should be now.

Maintaining high standards of freedom has never been easy. In the mid-1800's, for example, German immigrants in Fredericksburg, Texas, had lived in the United States fewer than ten years when the Civil war broke out, and though they had settled in a country which condoned slavery, they didn't support it. They kept to themselves, making a living and honoring their agreements with Native Americans when the Civil War forced their hand on the subject of slavery. Texas was the only state that put secession from the union to a popular vote, and Texans voted 2 to 1 to secede, but the immigrants in Fredericksburg had come to Texas for freedom, and they were not about to fight for any faction that would deny it to someone else. They were so adamant about personal freedom that they formed a military company and planned a march to Mexico, where they would board a ship to New York and join Union forces. They were massacred in Comfort, Texas, by the Confederates.

These were people who sacrificed everything for the cause of freedom. Texans regarded their strength of character so highly that, until recently, Comfort was home to the only Union monument in the Old Confederate south. It is inscribed in German.

Contrast their strength of character with Samantha Power's. She is our United Nations ambassador who won a Pulitzer prize for "A Problem from Hell: America and the Age of Genocide", a book in which she condemns the US government for turning a blind eye to 20th century genocide. In the book, she insinuates that advanced countries have both the right and the responsibility to "interfere" when atrocities occur. As stated in Wikipedia, her book looks at the "persistent failure of governments and the international community to collectively identify, recognize and then respond effectively to genocides." Really? To date, she has not made a single shrill speech about Islamic beheadings or prisoner flambeaus that would gather the media's attention. It turns out that she only talks a good game. If our ambassador, one who has made a name for herself by criticizing a lack of US intervention, can't make a weekly demand for exclusion of states which support slavery or abet the slave trade, we are *not* spreading freedom. We aren't even protecting against it, and neither is the United Nations.

"Slavery is a part of Islam. Slavery is part of jihad, and jihad will remain as long there is Islam."

Shaykh Saleh Al-Fawzan, 2003

That's a quote from a high-level Saudi jurist who represents a society which endorsed slavery until 1962. In my opinion, we should remove any who think like him from our country. Maintaining the sovereignty of our country's precepts may require that the United Nations be moved elsewhere, which would give the organization the opportunity and power to enrich more needy areas of the world. We don't have to leave the UN; our voice, if we still have one, needs to be heard. But think of the economic advantages of relocating the hub of international diplomacy to the birthplace of man! All those taxis, hotel rooms, and jet flights would build a sturdy economy for any country—theoretically, at least. Africa is the most appropriate place for the United Nations to live; learned debates on the subject of freedom and human rights might educate and improve the socio-economic welfare of African peoples. If Africa is truly the birthplace of man, it should also be the birthplace of the final fight against his enslavement.

Once, when I was about twelve years old, I went dove hunting with a friend. We were using his grandfather's 16 gauge single shot shotgun and my .22 rifle. Unaccompanied by adults, we were supposed to be taking turns shooting, but my friend was much bigger and stronger than I was. He would shoot that old 16 gauge and miss, and then I'd beg for my turn. Again and again, he shot and I begged, but I never got a turn to shoot. After at least a dozen misses, none of which produced so much as a feather, I'd had enough.

"I could do better than that with my .22!" I shouted, just as another dove-shaped missile whistled above us. Tipping my rifle in that direction, I pulled the trigger and hit it square in the breast. I think we killed a few more before we finished, because his grandmother cooked several for us that night, but I wore that story out.

Sometimes, no amount of begging works as well as a properly placed shot. To free the world from slavery for good, we should aim a shot at the Pan-African Parliament (PAP) by letting them know that the United States is intent on moving the United Nations headquarters to its new location in Africa. The PAP would be required to determine the location of the United Nations' new home within one year, then ten percent of American donations made to the UN would be sent to them out of the our budget for the U.N. IF they were unable to make a decision on where to host the U.N. that money would return to the United States Treasury and the free people from whom it was taken. At the point they designated a host country, the USA's grants for the United Nations would be garnisheed 10% and sent to the PAP in increasing annual increments for the next ten years. It would be cumulative, increasing by another 10% each year, so that all American donations to the UN would be sent to PAP by the end of the ten year period. At the end of those ten years, the United Nations would be living in the new Brasilia of Africa, and our subsequent contribution could be redirected to the UN at it's new home once more.

Some may laugh. I smile. In all likelihood, sending money to the PAP would be pouring money down a rat hole while the world watches. Letting the United Nations and the Pan-African Parliament determine where to build their new home and demanding that they adhere to Generally Accepted Accounting Practices would either generate change or end the United Nations for good. I don't see a downside to either result. In any case, pouring money into a sinkhole like the PAP couldn't be any worse than donating to the United Nations itself. Putting an end to an organization that effectively condones slavery is nothing but good. Relocating the complicit U.N. would be an invigorating booster vaccination for freedom.

"Let them be like the snail that

dissolves into slime..."

Psalms 58:8

When I was in high school, I had a friend who kept a most interesting couple of pets. Full grown, heavy, 6-foot rattlesnakes enjoyed the living room with the rest of his guests. Now, obviously, snakes will be snakes, so you can't just let them have the run of the house. He'd given them lip piercings, like the ones that many kids have today, but in this case the rings kept their mouths from opening completely. Their tongues could flick out and taste the world, but they couldn't bite.

Every four to six weeks he put them in their cage, removed their lip ring, and fed them a big fat rat. After that, they were happy again and smiling with their jewelry again. The most amazing thing about those snakes was how friendly they were when my friend took away their ability to be deadly. They would crawl around the living room and find a happy place to sun, they'd lay on your lap to be stroked like a lap dog, or they'd just crawl into a corner and coil up. They never tried to run away and hide in a closet. They liked Captain Kangaroo as much as the rest of us.

If they'd been introduced to you, they didn't even rattle unless something stirred them up. Slamming a door, dropping a glass, or the sound of action movies could put them on alert. An unfamiliar person in the room was a sure way to get them rattling, and a snake's rattle is a sound that makes people scream and instinctively back out of a room. Fast, and not infrequently on all fours. Sometimes sideways. Often in reverse.

The world we live in is corrupted by entities that have a bite much worse than any pit of rattlesnakes. According to news reports, large numbers of people in some countries like to meet on the streets every week and chant "Death to America." Often, the fact that Obama regards those countries as reliable negotiating partners causes me more than a bit of concern. At other times I surrender to the idea that we really aren't in control of events. The only things I can cling to constantly are my Bible and the gun that keeps the Islamic State out of my yard.

"For God has put it into their hearts to accomplish his purpose by agreeing to hand over to the

beast their royal authority, until God's words are fulfilled."

Revelation 17:17

I'm shocked—shocked, I tell you—that there are regimes and religions and areas of the world which still support slavery explicitly. Shocked.

There was a time when the British sponsored the West Africa Squadron to intercept and defeat the slave trade from Africa. Thousands of slave ships were captured, and hundreds of thousands of unfortunates were freed back to the continent of their birth. Wiki says that the British Empire devoted one sixth of its military resources to the West Africa Squadron, which put teeth in the British belief that *all men* have an intrinsic right to be free. If it weren't for British pressure, it is almost certain that the Saudis would *not* have banned slavery in 1962. Since the memory of slavery's evil is so fresh in Saudi culture, I wonder how much money they are spending to fight the scourge of its resurgence. There are living slave holders in Saudi Arabia. Don't you think it is time we heard the Saudi king, Protector of Two

Sanctuaries, tell the world that Islam does not support slavery?

The fact that free countries are not taking a stand against slavery leads me to think that we are standing on a dangerous precipice. Journalists have spilled gallons of ink on stories about people defacing and removing statues of famous Democrats who supported and fought a war to retain slavery. Re-enactments of marches against Democrat politician and policies during the Jim Crow era have led to jillions of bits being spewed across the internet. On the surface, America's rejection of slavery looks solid, but in recent history, the President of the United States—the first African American to hold that position—bowed to a Saudi king at one of their meetings. This strikes me as the greatest of all Pulitzer possibilities, but I must be the only one that sees it. The leader of the free world, apparently representing a segment of the world's population that suffered under slavery for hundreds of years, bowed to a Muslim king who most assuredly held slaves until he was almost forty. I would wager that he owned dozens of slaves. Scores, perhaps, including concubines who were minors. Probably dozens for each wife, too, and he had plenty of wives. King Abdullah's slaves, if they really were freed in 1962, might be in their 60's today. What would the surviving slaves of Saudi royalty have to say about the leader of the free world bowing to their former master? Their stories would probably make all those of Thomas Jefferson's ownership of slaves seem stale. No video though we have the chance! Funny that.

Our alliance with slave owners is just one of the issues that bother me. We will soon be living with more nuclear snakes in our midst, and our president has not pinned their lips. The increasing tension that Americans are feeling is much like the nation's anxiety over the Kansas-Nebraska Act of 1854. That piece of legislation negated the Missouri compromise and allowed a simple majority of white settlers "popular sovereignty," which meant that any new state could vote on whether or not to hold slaves. Standing against this travesty gave birth to the Republican Party. Today, Muslim immigrants want to bring slavery back into our world. Adhering to Sharia law, they disallow the education of women, and they insist on covering them up against their wishes. They throw acid on the faces of women who object to abuse. They have already demonstrated their readiness to excise a woman's clitoris and to strangle their own daughters while living in

Western nations. If cities become sanctuaries for them, as some have proposed, how will our society assimilate adherence to Sharia law? The meaning of "sanctuary" will be turned on its head if it defines a place where Muslim citizens forbid women to drive, to attend school, or to walk alone.

We can't allow it. The very idea should breathe fire into conservatives, because it mirrors the egregious environment that led to the founding of the Republican Party.

"From that day on, half of my men did the work, while the other half were equipped with spears, shields, bows and armor. The officers posted themselves behind all the people of Judah who were building the wall. Those who carried materials did their work with one hand and held a weapon in the other,"

Nehemiah 4:16-17

The wall of our democracy has been built with the will of the people and their right to express it at the voting box. The secret of our society's success is the Bill of Rights, which stands in the way of would-be slavers by protecting individuals' rights from collectivist oppression. Without debating whether or not ghost employees should be allowed to vote, or which party has the most expired voters, I think we could make serious headway as Republicans by addressing Voting Rights once again. It would appeal to millennials, it would be simple, and it is the right thing to do.

No law as important as the Voting Rights Act of 1965 can be limited to any particular region or make special exceptions for specific areas. The right to vote is foundational. Therefore, extend the Voting Rights Act of 1965 "special provisions" for every state! One additional extension should be the requirement to present official ID at voter registration and at polling locations. Voting, the most crucial function of any free government is certainly as important as writing a check at the grocery store. If American citizens are required to produce a valid ID when writing a check, then why should anyone object to being required to produce it at the voting booth? If our ability to choose our government officials is crushed by fraud, diluted by illegal voters with no interest in freedom, or polluted by those who wish to live under a different system of law that includes slavery, our country as we know it will not survive.

No one has the right to negate my votes or my Constitutional rights! Immigrants who place their own laws above our Constitution should not be allowed to enter the country. If we allow a large minority of slavers to filter into our society, we will eventually be forced into another horrific, purging ordeal. Civil War II.

> Item 8: "As freedom-loving Texans, we demand that our leaders expend whatever energy and funds necessary to protect our country and its borders. We also demand that our rights to the ballot be protected from all potential fraud, and our Constitution from any threat."

Although we live in a technologically advanced world, we are in danger of sliding into a dimly lit future. It is dizzying. Home computers are a relatively recent development in our history, but they are crazy efficient tools to spread the renewed lust for blood in the Muslim world. In the current decade, Afghanistan terrorists have shot girls in the head because they wanted to attend school. ISIS has beheaded prisoners. Women are sold on the internet. How is it possible for enlightened people to believe that perpetrators of such atrocities deserve any respect at all? Why don't the lofty minarets of Islamic thought rise up and demand that their societies address these horrors? Until they do, I won't be convinced that they don't approve of them.

Google up "honor killing." We must understand the world we live in—the world with which our government wants to share sovereignty. In my opinion, the Islamic culture disgraces every piece of history which has led to a civilized value for each individual, which regards individual rights as the foundation for sound government, and which has brought clarity to the division between civility and barbarism. If there is any group that deserves an honor killing, it is the Islamic world. Their Koranic rules corrode decency.

"But if you bite and devour one another, watch out that you are not consumed by one another"

Galatians 5:15

Perfected Preamble

(1.) As free Texans, we believe that no United States tax dollars may be invested, granted, loaned, or otherwise expended in any organization or country whose leader(s) do not endorse the First Amendment to the United States Constitution. We believe that this requirement should be clearly stated in the Constitution, and we request that our elected representative move with utmost urgency to introduce a Constitutional Amendment to that effect.

(2.) As informed and literate citizens, we are entitled to have our desires included as input during the construction of laws governing our society. Therefore, candidates for every office will actively represent our demand for a Constitutional amendment stating that no law can be put to a vote or signed by an executive until its complete and unchanged text has been released on the internet for seven (7) business days.

(3.) As intelligent citizens, Texans demand copyright in every cultural venture funded by taxpayers' dollars.

(4.) No elected Republican will vote "PRESENT" on any legislation and will do everything in their power to implement rules of law to disallow such votes.

(5.) Citizens of a free republic have the right to demand that their electoral decisions be honored immediately, and as such citizens, Texans demand that each elected candidate be sworn into office no later than twenty-four (24) hours after the Secretary of State certifies the results of that candidate's elective race.

(6.) Texans believe in self-reliance and demand that the government treat all income, earned or redistributed, with equal reporting requirements; toward that end and in order to spread fairness throughout our government, Texas introduces the 1099-GOV. In addition, no taxing entity may levy taxes without a biannual audit confirming their adherence to GAAP and Sarbanes-Oxley.

(7.) We, the people of Texas, deserve to know the workings of our government at all times and demand that all state bank account records be made public and immediately available on the internet.

(8.) As freedom-loving Texans, we demand that our leaders expend whatever energy and funds necessary to protect our country and its borders. We also demand that our rights to the ballot be protected from all potential fraud, and our Constitution from any threat.

If you believe in these preamble suggestions, it will be easy to get them passed at the precinct and county conventions. I'd be surprised if they drew much opposition because that is not where the real action lives. Please present them.

Then the three companies blew the trumpets and broke the pitchers-they held the torches in their left hands and the trumpets in their right hands for blowing-and they cried, "The sword of the Lord and of Gideon!"

Judges 7:20

Where to Now?

I've always run hot and inevitably cold on politics. Working for "the man" usually leaves me feeling used, and then I fall off the call list for a good long while until I find another reason to get involved. Still, I like to believe that this is a country where the people can make a difference, and if there was ever a time when that should be tested, this is it. With the Iranian bomb months away now, I feel an urgency that I haven't felt in decades.

The most immediate challenge we face is determining how to make an impact in the next election cycle. We need a plan. Without one we are nothing more than lone voices being run over by the Chamber of Commerce or whoever has bought the current crop of politicians. I know I sound like a broken record, but I really believe that if we base that plan on regaining personal freedom, we could grow loyal, unified adherents. We've got to do something organized about the havoc caused by liberals; they've divided Americans by class, race and income, and they've asked us to make nice with countries who believe in slavery. They've sent our money to the moon on a Russian rocket, and they seem bent on spending money that we don't have. Life on this trajectory seems finite and scary. We definitely need a plan.

A friend and I, raised up together at church, often visited each other's homes after church on Sundays, and we found adventures wherever they might be. We'd been in the nursery and Sunday school together, so we seemed to think similar thoughts at times, almost like sharing whatever brain there was between us. When either one of us ran across something intriguing, it was a given that we would investigate it together, even if it looked to have a serious downside. Fear and teenaged boys rarely meet outside disaster.

A short walk from my buddy's house was a really interesting yard just off Cave Lane and Nacogdoches. It drew the attention of teenagers in the neighborhood because it contained a sinkhole that was surrounded by a high chain-link fence topped with barbed wire. It appeared that the owner of the lot used the sinkhole as a dump for lawn debris. From rumor (and then experience), we boys knew there was a cave at the bottom, and we moved the debris around to enable passage to the cave. Our goal was to keep the entrance available but camouflaged, because if we were noticed by adults, bad things would happen. They said.

Climbing a tree and then dropping into the enclosure was trivial. The passage to the cave was a boy-sized hole descending vertically for about ten feet, and we had to enter it head down in order to scoot into the first little chamber. Until I headed straight down into the bowels of the earth, my descent slowed only by a water hose tied to a log above and the spreading of my legs, I couldn't have imagined how it felt. I wouldn't call it fear, exactly, but playing snake is not in a man's character without learning it. Who ate the first oyster?

Once my friend and I were down into the cave, we explored. After going through the first two chambers, we could almost stand. As we continued further in, the holes in the ground got really big. We never found any interesting cave formations. It was more like a water flush for the neighborhood, with rock walls and mud floors tracing underneath San Antonio, trending, we believed, toward the Alamo. That was our story, anyway, and every trip we took into the cave seemed to reinforce the idea. In fact, on one of our deepest treks, we found numbered stakes in the walls, and the numbers got smaller as we went deeper.

Before each trip into the cave, we planned. Check the weather. We didn't want to get flushed. Flashlights and extra batteries. A ball of string in case we went in a new area. A small can of spray paint. Real experts we perceived we was. Those were the alert days. The thing we never thought to do was to leave a note saying to look for us there if we went missing.

Airline pilots have a checklist for a reason: humans get lackadaisical. It is just our nature. Without one, some of us perfect Stupid as we go.

On one particular day it was a great morning, and my buddy and I just couldn't wait to go down in the cave. We didn't cross off the checklist, and we slid down the passage with a single flashlight. We were making serious headway into the cave as we transited areas that we knew like the backs of our hands (pride), and then we arrived in new area where the passage opened into a giant crack, maybe six feet wide and twenty feet tall. It crossed our passage at the perpendicular, and it was multi-leveled: other passages opened from it at every angle and height.

My friend lowered himself down to the floor level, maybe ten feet below me. Reaching out, I dropped the flashlight! Both of us watched as it fell to the floor near my buddy's feet without a sound. It hit the ground and blinked out as it separated into parts, and we both said, "Oh, shit!" at the same time.

There's dark, and then there's underground dark that ain't never seen a photon. During the minutes that followed, we felt around for the parts of our only flashlight, put it back together by feel, prayed with feeling, and realized the folly of our ways. I'd like to say that this was the last time that either of us failed to plan, but.

I'd like to say that freedom is spreading through the world, too, and that being a fully engaged citizen isn't necessary because the Constitution is all we need to protect our freedom. That statement is as enlightened as cave exploration without a flashlight. Billions of people who neither know or understand freedom are moving around the world—they're calling it "migrating" this week—and if we fail to recommit ourselves to our legacy, our children will be condemned to the influence of barbarism. If immigrants to our shores bring anti-freedom philosophies, we have no choice but to insist that the philosophy of freedom takes precedence within our borders. As Star Trek's Captain Kirk said when his right to judge barbarity was called into question, "Who do I have to be?"

We need to face the mongrel hordes head-on with freedom as our standard. They may hold up signs saying "To Hell with Freedom," but I come from the heritage of the Alamo.

"Your word is a lamp for my feet, a light on my path."

Psalms 119:105

Any good battle plan is made up of several parts, which include obfuscation, redirection, and direct action.

Obfuscation

One of the most obvious things that WE Republicans have had to deal with over the last few elections is candidates who lie. "When I get to Washington, I'll…." (*Do nothing.*) It has made my last nerve raw.

"Make America Great Again." Will any Republican candidate disagree with this statement? I hope not. Even the most cynical RINO's surely believe that the goal we all seek is crystallized in that phrase. We don't need to measure candidates' support for it: all I'm going to suggest is that we make them wear it.

At every Republican party meeting, fundraiser, or cocktail party, candidates and their supporters should wear a "Make America Great Again" hat or pin. It may have been introduced by Donald Trump, but it is the idea that finally penetrates the body politic. We don't have to be Trump supporters to believe that it is a worthy goal. Even being "Ready for Hillary" shouldn't prevent anyone from believing that it is good to Make America Great Again.

"Are you a Trump supporter?" people will ask.

I would reply in one of several ways (take your pick): "Do you like my hat?" or "Don't you believe America can be great again?" Maybe, "Don't you think America was greater before?" or "Where do you believe we're heading?".

The only purpose for your allowing this conversation, if that's what it is, is to address *their* beliefs on whether America ever was, or could be, great again. Put them on the defensive and don't let up. This is not really a political issue, is it? If they don't believe in what America was intended to be, trying to convince them is a waste of time. Our vision is clear, defined in the Constitution, and unworthy of discussion with those who believe we are Greece. Casting our pearls before swine, so to speak, offers little satisfaction.

Wear the MAGA hats at the Precinct conventions. Wear them at the County Conventions. Wear them all around the state convention location, but *never* in any formal state convention meeting or on the convention floor. We must wave the flag but we can't allow ourselves to be a target when within range of the elite state convention crews. If every restaurant around the convention site is filled with "Make America Great Again" hats and buttons, but they see none on the convention floor, it will make them all squirm.

Redirection

The problem with money in politics is that candidates really do try to buy votes with it. I ran from funding the Republican Party directly when I discovered that, after an election, leadership opposed issues they had claimed to support. Then I decided that I'd send money to individual candidates whom I thought best measured up to my goals. I observed that this approach led to candidates being totally immune from party discipline. They lost the big picture and literally became owned by big donors. I would like to throw them all out now, with very few exceptions.

In order to register our disgust up and down

the line, and to let the Republican Party know that they've approached the end of our tolerance for being ignored, I suggest that we make all donations in a rather unique way. I think we should send our checks to the state organization so that the payee is "Candidate Joe C/O RPOT," and follow up with a personal note to the candidate as a notification of the contribution. (The last time I checked, the Republican Party of Texas lockbox account was found at: Republican Party of Texas, P.O. Box 2206, Austin, TX 78768.) We would no doubt get lots of push-back from the candidates. They'll tell us how hard it is to cash a check that way. Walking a check like that to a bank teller might lead to some questioning, but in the end, the bank would most likely allow the deposit.

When a check is sent to the state organization, it goes to a lockbox account and is deposited without issue. By making the check out to the Candidate Joe C/O RPOT, Joe is forced to beat his way back to the party to get the money. The state party also becomes aware that something is out of the ordinary.

In the Romney-Obama election, more than 8 million Republican voters stayed home. They failed to register a real opinion. In effect, they voted "Present." I suspect that, like me, they will never vote Democrat because they are fearful for our country. They are also, like me, finding it harder and harder to vote Republican. "CO" checks would alert the Republican Party to the fact that large numbers of us are moving away from them and their status quo.

The only other nationally registered option is to vote Libertarian. At this point in time, a Libertarian vote would be a true protest vote, and when we think about it, what damage could a Libertarian congressman or senator do that isn't already being done by either main stream party? Though the workings of government would make their strong stance tough to maintain, the Libertarians' political philosophy is superior to that of the Democrats, and certainly more American. While Democrats believe in open borders, Libertarians would never allow 100,000 men in Wehrmacht uniforms to cross our borders without being fingerprinted or taking an oath of loyalty to the Constitution. That alone would make them a better choice than the Democrats. (But who are we kidding? Democrats, Republicans, and up until the early fall of 2015, every European party was fully prepared to let millions of strangers into their homelands.) Democrats have already been caught arming drug cartels in Mexico, but at least the Libertarians believe that American citizens should be armed, too. Again, better than the Democrats. Whether the Libertarians' philosophy meshes with ours or not, their most important stance is the belief that the United States Constitution is the supreme law of the land. In contrast, at no point in the arc of Obama's career has he actually believed in the Constitution. In fact, he has stated that it was improperly written.

Migration issues raise a point that no one seems to address: what about the documentation of those who cross the border without visas? Is it wrong to ask these people where they came from? Surely not, and if it isn't wrong to ask where they came from, it is certainly reasonable to inquire whether or not they support the First Amendment. I'd like to think we can agree that we shouldn't let people into the country who are opposed to our fundamental freedoms, especially when political divisions between countries are getting fuzzier than our borders. A case in point: Mexico is working to get their citizens the right to vote in our country. Really.

It would be of interest to me to poll immigrants about their home countries when they fill out their entry forms. Maybe we should ask questions about the governments they have been eager to leave. We could ask if they would support changing the Constitution of, say, Syria to include the First Amendment. If they've actually been caught, I'd like them to have fingerprints like I got at the DMV.

I'm down a rabbit hole again, distracted by freedom. Sorry.

Direct Action

We take direct action when we enter the Republican Party of Texas state convention and actually fight for our ideas. Hopefully, at this stage, the revised preamble has been submitted by at least one county convention. Typically, the elite Republicans simply drop ideas that they don't favor. Unless someone ramrods a new resolution through the temporary committees during the week before the convention, a group's elected concerns may, literally, end up on the ground.

In the most hopeful of worlds, the preamble that I've laid out will have been submitted by multiple county conventions. At that point, it would surely get a hearing…? In any case, it is important that someone shepherd any resolution to replace the preamble so that it can actually get to the convention floor. This means that someone will have to be at the convention site during the week preceding the convention.

There is no step along the way where it can *not* be destroyed by a lack of attention.

The Temporary Platform Committee has three people chosen from the members of the State Republican Executive Committee who can cast secret ballots preventing you from knowing how much influence you had. With those secret ballots, they can kill *any* idea, even if it has popular support during the discussion time reserved for it and grassroots support across the state. They can kill it if they simply don't like it. If they kill it, it's pretty much done unless someone is willing to bring it up at the full committee. This is very rare.

From this point on, we're talking about the actual state convention. I won't waste time on a detailed description of what happens because I've always been surprised when I tried to do something there. Still, I have a couple of suggestions:

   1. Have a pocket copy of Roberts Rules of Order on your person, and be familiar with it.

   2. Be especially familiar with "Point of Order" and "Challenge the Ruling of the Chair."

   3. Work with a cell group. Find a couple of like-minded people and plan

to be on the floor together. Three is
the perfect number. Two people stand
in the line to speak as they are
recognized by the chair. The third
person brings food and drink and holds
the line for restroom breaks. I have
spent an entire day in queue and not
had the chance to have my say.

As free people, we must use the legal tools at
our disposal to effect change. Unfortunately, I
have no idea what to expect. I do know that
we've been losing for decades.

Years ago, my Boy Scout troop went
camping at Camp Bandina, named for its
location on the Medina River between Bandera
and Medina. Many rivers in Texas are what
people in other states call streams or creeks.
Where rivers are concerned, Texans use big
words for little things because water is precious
to us, and the Medina River is one of those
little things. It's about 10 yards wide for most
of its course, and it is lined with giant bald
cypress trees that two people can't reach
around.

In most places its depth varies from ankle-
deep to barely above an average man's head. In
the place where we camped, though, there is a
little half-moon waterfall about two feet high,
and over a period of millennia, its curved shape
has carved out a sink hole almost twenty feet

deep. The curve also generates an energetic whirlpool in the water. Floating debris tends to hang around in the center of the whirlpool, displaying evidence of recent rises in the water level. At the time, one of the cypress trees grew over the pool at the perfect angle for an easy ascent to a platform that had been built by someone with a vision for fun, and from the platform we could catch a rope and swing out over the pool, gaining enough altitude to do an easy flip into the water. An experienced rope swinger could probably have done a two-and-a-half, but that wasn't me.

The young boys in our group were not on the diving team. We would swing out over the water and jump into it, but we were not particularly daring. And once again, I was daring beyond my ability, and certainly beyond common sense. Now, everyone has heard about the danger of diving into unknown waters, but this was known water. I'd been into this pool hundreds of times, and I knew where the bottom was and where the cypress roots lived. Every time we went in the water, we all believed that we might noodle out some giant catfish, as if we knew about that, too. In short, I had probably put my hands on every piece of that sink hole at one time or another.

I dove into the water. The dive, I feel sure,

was absolutely perfect because my angle of entry seemed just right for slipping into the water. After entering the water, I gracefully curved myself toward the surface to avoid hitting the bottom. The second or two of personal pride in my diving skill was short circuited by the first sensation of something hard at my fingertips. At almost the exact moment my brain registered that things were not going as planned, I was jammed waste deep in the mud between some cypress roots. Jammed up hard.

Once more I was stuck in a world without photons. My legs were completely free and waving madly in the current of the whirlpool, but I couldn't even move my arms to my side. I hadn't expected or planned for this, and black fear washed all rational thought from my mind. (It is because of this experience that I respect Navy SEALS so much. They are trained to work in places like this and actually perform delicate and dangerous tasks.

I did have enough sense to not scream under water, but it was a minute or so before I heard the first splashes of my rescuers. They grabbed hold of my legs and extracted me just seconds to spare. Once I was safe, we had quite a laugh. I would like to say I learned something from that, but the only apparent lasting effect is that I can sense when I'm flailing.

We are entering a very dangerous time for our country and our culture. Questions which we cannot even imagine now will be asked. If you don't sense that the Constitutional framework of our country is flailing, I don't think you're paying attention.

If a swarm of unarmed young men invades our country, do we have the ability as Americans to preserve and enforce our beliefs, or will we compromise with whatever crosses our borders? How many people can camp in your living room without leaving their mark? Do you believe we have the right and the duty to defend our culture, our Constitution, and our homes? Hard questions and hard times are heading our way. We'd better prepare ourselves for both, think hard about our path, and cling to our Constitution. If we don't, we'll lose everything that the United States stands for.

Therefore I run thus: not with uncertainty. Thus I fight: not as one who beats the air.

1 Corinthians 9:26

"…we warn and adjure earnestly . . . that no one in the future dare to vex anyone, despoil him of his possessions, reduce him to servitude, **or lend aid and favor to those who give themselves up to these practices**, or exercise that inhuman traffic by which the Blacks, as if they were not men but rather animals, having been brought into servitude, in no matter what way, are, without any distinction, in contempt of the rights of justice and humanity, bought, sold, and devoted sometimes to the hardest labor."

Pope Gregory XVI, 1839

PBS Data Structures

In this section I'll lay out the data library required to effect the task of the Crowd-sourced FOIA system that I mentioned in the Public Bank Statements chapter. In any table-driven database system, a developer usually starts building the tables from the most obvious and simple and gradually moves to the hard stuff. For example, I know I'm going to need a 'person' table with name and address, and once I get that, I can move on to more complex parts of the system. Here, I'm going to start at the point where it gets interesting and leave all the foundational parts at the end for people who are really interested. And, I'm going to forget to include plenty, I'm sure. Sorry.

This is a critical part, I think, of taking back our government. We can't allow our leaders to continue hiding their actions from us any longer. We can no longer watch the game in which politicians make a public show of activity on our behalf, but behind the scenes plan our futures with secret communications and treaties. We have a right to know whether or not their agendas match our desires, and exposing their activity to sunshine is the only way we'll find out.

One of the places where we can view government making some effort is www.data.gov. Browsing this site reveals that there are plenty of government reports, but practically nothing that deals with government spending. Politicians are attempting to bury us in reports... once again. If you don't believe that, enter "American Express" in the search field to see what they want you to review. Even if spending information were available, there is no application which coordinates the ability of citizens to focus their attention on a particular dataset. Managing a set of people who really want to make government work and who are willing to analyze the data of government...that tool is not here. Yet.

Note that the "hottag" is a flag used to designate an item in a tranche of documents/statements that the crowd has found interesting. A hot item is something that someone might want to review further. When enough people inside the system comment on an available document, statement or report as interesting, it becomes a hot item . Once the item is hot, it is visible from several different layers so that WE can see exactly where interesting things are happening. All users could then invite their review and add additional comments. In other words, as we're all working together, independently, those things that are interesting float to the surface and become MORE visible as people remark on them.

Below are some of the tables that I think would make a real system. They don't define a complete system or the best system, but they would be a start. Literally thousands of systems that manage comments already exist, so there is no sense trying to re-invent that wheel.

Statement Header Table

In order to demonstrate part of the crowd-sourcing application, this first table can viewed as the description of the data collection: it could be the description, for example, of bank or credit card statements, documents requested via FOIA, or a particular tranche released on a certain day. We are all familiar enough with bank statements to know that they can clutter a desk. Ordering them by month would be a typical method of looking at a given data set. In the table below, I've given limited example data, but enough to explain this step of the process.

The interesting thing about this record is that it contains the "reviewed" field, which is represented in my mind as a pie chart or a similar graphic associated the with the reviewed status of each bank statement. A quick glance across the screen would let everyone know which statements have not been completely reviewed. In like manner, if there are particular transactions in that statement that people have flagged as hot items, they would be obvious, also.

Certainly there are at least one (1) million people who love this country enough to spend a few minutes reviewing documents in the hope that they might—just might—help regain control of government. If we can't work together to pick up the broken pieces that our leaders have left us, it could be a millennial quest to return freedom to our progeny.

One million pairs of civilian eyeballs focused like a laser beam on the operations of government. Sounds great, doesn't it?

```
1    CREATE TABLE statement_header (
2      entity_id  INT UNSIGNED NOT NULL,
3      acct_id INT UNSIGNED NOT NULL,
4      acct_type  INT  UNSIGNED NOT NULL,            # 1=bank stmtt, 2 = credit stmt, 3= doc tranche
5      src_document VARCHAR(50),                      # link to source document (PDF?)
6      statement_desc VARCHAR(50),                    # eg. "Wells Fargo xxxxx3124"
7      statement_id INT UNSIGNED AUTO_INCREMENT,      # direct identifier this statement/tranche
8
9      statement_detail_seq INT UNSIGNED,             # increment as detail line items added
10     disp_desc VARCHAR(30),                         # displayed in the title above all the items, eg.
11     from_date DATE,                                # eg. October 1, 2015
12     to_date DATE,                                  # eg. November 1, 2015
13     audit_num INT UNSIGNED NOT NULL DEFAULT 1,     # num records when 'crowd-source' button hit
14     num_detail_items INT UNSIGNED NOT NULL,        # number items under header
15     last_sourced INT UNSIGNED NOT NULL,            # last transaction crowd-sourced
16        /* every time the 'crowd source' button is hit a certain 'audit_nm' of items are returned and when the
17     end of the item list is hit it starts over again at '1'.  In this way, it just rotates through continually
18     allowing focused review of All THE items.
19     */
20     rollover INT UNSIGNED,                         # counts each pass though data set
21     last_comment TIMESTAMP,
22     last_comment_opid VARCHAR(20),                 # last commenter opid
23     last_comment_transaction INT,                  # last transaction crowd-sourced
24     reviewed INT UNSIGNED,                         # percent crowd-source pie chart
25     cnt_comments INT UNSIGNED,                     # total comments this statement
26     memo VARCHAR(512),
27     hottag BOOL,                                   # any doc/image/etc tagged as hot
28     hottag_opid  VARCHAR(30),
29     hottag_date TIMESTAMP,
30     edit_opid VARCHAR(30),                         # person opid that add/chg record
31     edit_time  TIMESTAMP,                          # date/time of opid edit
32     CONSTRAINT pk_itemhd PRIMARY KEY (statement_id)
33   );
```

"For if they fall, one will lift up his companion. But woe to him who is alone when he falls, For he has no one to help him up."

Ecclesiastes 4:10

Statement Detail Table

This table represents the actual line items to be crowd-sourced, whether they are individual documents in a large tranche or individual lines in a monthly credit card statement. Obviously, some fields have no purpose in the beginning, and some fields don't appear because it's all a big guess until we see actual data.

There will be a "Crowd Source" button available on the screen to anyone reviewing a statement, and when the reviewer clicks the button, a subset of the transactions is returned for review. When this happens, the reviewer knows that they are reviewing things that are focusing attention on areas that need review. I keep coming back to this because it is so very interesting and almost trivial compared to what gaming sites have to manage. If 7.3 million people logged into the site on one day to crowd-source the CIA document tranche, it would be analyzed in ONE DAY. That's fewer people than some online gaming sites have as monthly subscriptions so I know it is a reality(soon). Via hot buttons, their attention could also be given to items that have been flagged by others, and it that way users are alerted to what has already captured people's attention.

We're focusing internet interest to cover all the tentacles of government. The goal is to have data loaded just like Quicken does: ping the bank, get the data file, and load it. If the banks can get your personal accounts to load Quicken, you know they can do it for OUR government accounts to load our monitoring system. When it comes to things like huge tranches of documents from the CIA, the State Department, the IRS or any government official that maintained separate data libraries, we'll just have to futz around with it until we get it right.

While researching, I found that today's cost of scanning from paper is eight cents per page. Hopefully, we've learned from the former Secretary of State that electronic government records don't have to be turned to paper before they are released to The People. The trees will thank us.

This is the part of the system that supplies the really interesting features. Being able to see what has been reviewed and how often, and whether it has gathered enough attention to be flagged as a HOT item.

```
1    CREATE TABLE statement_details (
2    entity_id INT UNSIGNED NOT NULL,
3    acct_id  INT UNSIGNED NOT NULL,
4    acct_type  INT  UNSIGNED NOT NULL,              # 1= bank stmt, 2= credit stmt, 3=document
5    statement_id INT UNSIGNED,                       # from statement/tranche header
6    statement_detail_id INT UNSIGNED NOT NULL AUTO_INCREMENT,      # global sequence number
7    trans_date DATE,                                 # transaction date
8    trans_clear_date DATE,                           # date it clearned bank/credit card if relevant
9    page_ref  INT UNSIGNED,                          # If it refers to a page on a pdf or something
10   comment_seq INT UNSIGNED NOT NULL,               # pull num to sequence the comments
11   payee VARCHAR(50),
12   payee_id INT UNSIGNED,
13   amount  DECIMAL(10,2),                           # government CAN spend billions!
14   bank_memo VARCHAR(512),                          # any memo associated with this from the bank
15   system_memo VARCHAR(512),                        # system administrator for whatever reason
16   system_memo_opid VARCHAR(30),                    # opid of sysadmin the created system memo
17   system_memo_date  TIMESTAMP,
18   comments BOOL,                                   # comments can be turned off on an
19   comments_cnt INT UNSIGNED,                       # each comment increments this; sets  font
20   verified BOOL,                                   # authority said this can be released to pubic
21   verified_opid VARCHAR(30),
22   verified_date TIMESTAMP,
23   hottag BOOL,                                     # this tagged hot item, cascades to headers
24   hottag_opid   VARCHAR(30),
25   hottag_date TIMESTAMP,
26   edit_opid VARCHAR(30),                           # person that add/chg record
27   edit_time  TIMESTAMP,                            # date/time of opid edit
28   CONSTRAINT pk_statement_detail PRIMARY KEY (statement_detail_id)
29   );
```

County Table

Seeing as how WE Republicans are from Texas in this case, this system will be focused mainly on retrieving data from the city, county and state files. These foundational database tables don't change much, if ever, which is nice.

Of course, the county table will be used to track a parish, if this ever grows out of state.

```
1    CREATE TABLE county(
2    county_name VARCHAR(30) NOT NULL,
3    state_id CHAR(2) NOT NULL DEFAULT 'TX',        # of course it is starting HERE
4    county_id INT UNSIGNED NOT NULL AUTO_INCREMENT,
5    fips VARCHAR(5) NULL,                          /* Federal Information Processing Code
6      that uniquely identifies counties ("FIPS 6-4") or county equivalents eq parish
7    */
8    county_seat VARCHAR(60) NULL,
9    area FLOAT NULL,
10   population FLOAT NULL,
11   memo VARCHAR(512) NULL,
12   website VARCHAR(128),                          # url county website
13   edit_opid VARCHAR(30),                         # person that add/chg record
14   edit_time TIMESTAMP,                           # date/time of opid edit
15   CONSTRAINT pk_county PRIMARY KEY (county_id)
16   );
```

City Table

The city table is pretty self-explanatory. The books of a large city are going to reveal some real treasures, I'm sure.

```
CREATE TABLE city (
state_id CHAR(2) NOT NULL DEFAULT 'TX',
city_nm VARCHAR(30) NOT NULL,
city_id INT UNSIGNED NOT NULL AUTO_INCREMENT,
fips VARCHAR(8),                           # FIPS for city eg. "48-11464"
population FLOAT NULL,
area FLOAT NULL,
city_memo VARCHAR(128),
counties VARCHAR(50),                      # cities span counties often
city_website VARCHAR(128),
edit_opid VARCHAR(30),                     # person that add/chg record
edit_time TIMESTAMP,                       # date/time of opid edit
CONSTRAINT pk_city PRIMARY KEY (city_id)
);
```

Elective Entity

This table is keeps track of the organization which supplies the documents for review. Thousands of such organizations exist in Texas. If we demand FOIA access at the smaller ones, we can sneak up on the bigger ones. Starting at lower levels of government offers an advantage: they have fewer resources to jerk honest citizens around than does the 800-pound gorilla (the state).

```
1    CREATE TABLE elective_entity (
2    entity_name VARCHAR(50) NOT NULL,              # eg. "Railroad Commission of Texas"
3    entity_id       INT UNSIGNED NOT NULL AUTO_INCREMENT,
4    entity_state  CHAR(2) NULL,    # some entities may not be tied to a state, leave room for Feds
5    county_id INT UNSIGNED NOT NULL,
6    city_id INT UNSIGNED NULL,                      # LCRA isn't really city associated
7    addr  VARCHAR(128),
8    memo VARCHAR(128),
9    entity_website VARCHAR(128)
10   hottag BOOL,                                    # any doc/image/etc has been tagged as hot
11   hottag_opid  VARCHAR(30),
12   hottag_date TIMESTAMP,
13   edit_opid VARCHAR(30),                          # person that add/chg record
14   edit_time  TIMESTAMP,                           # date/time of opid edit
15   CONSTRAINT pk_elective PRIMARY KEY (entity_id)
16   );
17
```

Elected People

The 'elected_p' table tracks the people (p's) who fill positions in government organizations. In my opinion, one of the problems we have in government is the inability to stamp a single thing with somebody's name. The buck can always be passed. Do we even today, years later, know who gave Secretary of State Clinton permission to keep a private email server?

```
CREATE TABLE elected_p (
entity_id INT NOT NULL,
title VARCHAR(128),                      # eg. Assistant Director Issues for  Texas Waterways
position_id INT UNSIGNED NULL,           # eg.  Chairman/Assistant
person_id INT UNSIGNED NOT NULL,
first_elected DATE,                      # date/time first elected/appointed
last_elected  DATE,                      # date last elected/appointed
appointed  DATE,                         #  some positions are appointed
term_end  DATE,
political_party  VARCHAR(20),
personal_img  BLOB,                      # every politico has a head shot
memo_cv  BLOb,                           # should we save their resume? Or link wiki's?
media_bar VARCHAR(512),                  # link their YouTubes
cell_ph  VARCHAR(12),                    # cell phone
work_ph  VARCHAR(30),                    # work phone
email VARCHAR(30),
website VARCHAR(50),                     # their link, to Clinton Global Initiative?\
image_border INT UNSIGNED,               # decorate Republicans different than demos, libs
image_footer INT UNSIGNED,
edit_opid VARCHAR(30),                   # person that add/chg record
edit_time  TIMESTAMP,                    # date/time of opid edit
CONSTRAINT pk_position PRIMARY KEY (position_id)
);
```

Line Item Fonts

I think this area is really cute. As the number of comments about a particular statement, line item or detail grows, its appearance on the screen changes. As a growing number of comments are applied to a document/statement detail item, this table configures how it is viewed. It won't be configured as a hot item until designated in some way by a system process which is yet to be determined.

I'm sure there are many more ways to clue people into something interesting happening but I really like the idea of being able to glance across a screen full of information and have the ones that have been interrogated a bunch of times be blinking. In red?

```
CREATE TABLE line_item_fonts (
font_id INT UNSIGNED NOT NULL AUTO_INCREMENT,
comments_threshold  INT UNSIGNED,              # font to display item  if this many comments
font_size INT UNSIGNED,
font_color INT UNSIGNED,
edit_opid VARCHAR(30),                         # person that add/chg record
edit_time  TIMESTAMP,                          # date/time of opid edit
CONSTRAINT pk_fonts PRIMARY KEY (font_id)
);
```

Story Listing

SIMMERING FEAR

Tweeeeeeeth.......................................19

Bumblebee Alley.................................22

Bull Rider...24

Movie Madness...................................28

Sabinal tenderizer...............................31

DISS-SATIFIED

Oral boot camp tags............................45

Smoking GLO.....................................53

Cutting bulls.......................................57

FIRST THE FIRST

Trophy Sorting....................................71

Betting on Hillary...............................75

NO MORE SAUSAGE

Twitch off...95

Sweet Sausage....................................98

PUBLIC BROADCASTING

Skidding at It.....................................108

VOTE YOUR MIND

Bat Geometry..................................116

VOTES CHANGE LEADERSHIP

Cattle Drive 35.............................119

Cotton Pickin...............................131

Testing the buzz...........................135

1099-GOV

Active Snooze...............................146

Cover me......................................154

Strikes off....................................166

YOUR Government Records

Chile for beans.............................178

Weak TEA...................................185

United States Sovereignty

Touched by no angel.....................198

Trick Shot....................................210

Snake bit......................................213

Where to Now?

No Photons...................................226

Half free.......................................238

Author Pitch

I've spent enough time talking about myself. Crazy, stupid, silly stories about my past have already told you nearly everything. The stories I've shared are deeper in me than most that I've seen in artists' bios about their motivations and history. I have more, to be sure, but I have to hold them close because I want to address the questions they raise at home. I won't be ready to share some of them until my kids are forty.

You certainly understand by now why I don't consider myself a writer, but I was—and still am—inflamed by Obama, and I had to express myself somehow. The same goes for my graphic arts, such as they are.

My paintings are listed at SenatorMark4.ArtistWebSites.com and I would really appreciate your visiting and purchasing something. The LOL painting is 5 feet x 6 feet, and was driven by my fear for the country's future. When a rank beginner looks at a blank canvas, he has to be seriously motivated, or anxious, to finish something the size of a barn door. If you go to my website, you can zoom in on the details. They will make you laugh. Personally, I love the fact Obama's belt buckle is a "57". The tote bags are especially interesting because they work to hold your possessions safely as opposed to what Obama's administration is working to accomplish.

It wasn't until I got into this that I realized how much the left is sponsoring the upstream pollution that society is imbibing today. If the picture of a urine crucifix can fetch $200,000, I think there should be some conservative who would like to hand out handbags based on the LOL painting. "(candidate name) OR more..." emblazoned at the top would replace 10,000 words. I would be very pleased to see them flying out of my art site!

The history of my scribblings can be found on Amazon in "Christia: A Screenplay, which focuses on the African-American experience in West Texas. I thought that a story containing West Texas, cowboys, robots, and a sprinkling of Texas history would do better. The other one is 21st Century Homestead, which outlines a new idea for a self-sufficient urban farm. It's a short little pamphlet which is capped off by another Marky story, another laughing matter which documents an event which is rare to the extreme.

Obviously, my family needs support to protect me from myself. Please help.

FINIS CORONAT OPUS